Life Testing
and
Reliability Estimation

S. K. Sinha
Department of Statistics
University of Manitoba
Winnipeg, Canada

B. K. Kale
Department of Statistics
University of Poona
Poona, India

A HALSTED PRESS BOOK

JOHN WILEY & SONS
New York   London   Sydney   Toronto

Copyright © 1980, WILEY EASTERN LIMITED, New Delhi

Published in the U.S.A., Canada and
Latin America by Halsted Press,
a Division of John Wiley & Sons, Inc., New York

Library of Congress Cataloging in Publication Data

Sinha, Snehesh Kumar.
    Life testing and reliability estimation.

    "A Halsted Press book."
    Includes bibliographical references.
    1. Failure time data analysis.   2. Reliability (Engineering)
I. Kale, Balvant Keshav, 1933–   joint author.   II. Title
QA276.S53   519.5   80-14971
ISBN 0-470-26911-1

Printed in India at Urvashi Press, Meerut.

To
Rubena, Pamela and Debashis Sinha
and
Sushila, Aniruddha and Meena Kale

Preface

Life testing is an important and useful section in the area of statistics for engineering sciences. This book is an attempt to present the theory and methodology of inference, mainly the estimation and tests of hypotheses problems, in this area. In the last few years, several papers have appeared in a variety of journals and a few years ago one of the authors, Professor Sinha, started a first year graduate course on this topic at the University of Manitoba. This text is based on the material collected for the course and presented to the students over the last four years. It is assumed that the students have two years of calculus background and a course in mathematical statistics including standard distribution theory and introduction to Estimation and Tests of Hypotheses. This book is designed to serve as an introductory text for a one semester course for students of Statistics as well as Engineering and Mathematics.

After an introductory chapter on a variety of models used to describe the failure time distributions, we consider in Chapter 1, the Exponential model in some depth. In Life Testing experiments the Exponential model has played a role very similar to that played by the normal distribution in agricultural experiments. Chapters 2 and 3 consider the models based on Gamma, Normal and other related distributions. Chapter 4 covers the Mixture and Compound distributions. In all these chapters we consider estimation problems only, as these arise in complete samples as well as the failure and time censored samples. The methods of estimation considered are (i) Maximum Likelihood (MLE), (ii) Uniformly Minimum Variance Unbiased (UMVUE), (iii) Method of Moments, and (iv) the Linear Combination of Order Statistics of the sample. The UMVUE approach requires the completeness property which in turn has increased the level of the material presented. We, however, think that the UMVUE approach is one of the basic methods particularly in the Engineering context, since many of the estimation problems, such as filtering or signal detection, use the UMVUE approach. Chapter 5 deals with the Tests of Hypotheses and Confidence Intervals for the parameters occurring in the models, as well as the reliability function. The approach here is purely decision theoretic based on the Neyman-Pearson theory. Chapter 6 deals with the Bayesian approach and Chapter 7 deals with Estimation of the Reliability Function for the series as well as the parallel systems. A number of illustrative examples have been worked out at the end of each section and exercises are given at the end of each chapter. Wherever possible, we have tried to use real life data. When such data are not available, at least for inclusion in the text, we have tried to illustrate the techniques by using simulated data. There are two appendices, one outlining the basic Distribution Theory used and the other outlining the basic ideas of Estimation and Tests of Hypotheses. The material available is very vast and we had to choose the material selectively but we hope that what we have selected serves as a good introduction and encourages the reader to go further in this interesting field.

We wish to thank the authorities of the Universities of Manitoba and Poona as well as the University of Ilorin, Nigeria for providing excellent support facilities. We also thank the National Science and Engineering Research Council of Canada for the financial support without which the extensive amount of typing and duplicating needed would have been impossible. Typing and retyping several drafts was a demanding task and we owe much for the patient and understanding secretarial help given to us by Mrs. E. Loewen, Sharon McKinley, and Mrs. Sabane. Thanks are also due to the referee whose valuable suggestions have gone a long way in improving the text.

July, 1979   S. K. SINHA
Winnipeg, Manitoba, Canada   B. K. KALE

Contents

Introduction

When we buy a transistor radio or a scooter or even a simple product such as an electric bulb or a battery we expect it to function properly for a reasonable period of time. When a manufacturer floats a new brand of light bulb in the market, he would like his customers to have some information about the average life of his product. Life testing experiments are designed to measure the average life of the component or to answer such questions as 'what is the probability that the item will fail in the time interval $[t_0, t_0 + t]$ given that it was working at time t_0'.

In a simple life testing experiment a number of items are subjected to tests and the data consist of the recorded lives of all or some of the items. No matter how efficient the manufacturing process is, one or more failures may occur. This failure may be due to:

(i) careless planning, substandard equipment and raw material used, lack of proper quality control, etc.;
(ii) random or chance causes. Random failures occur quite unpredictably at random intervals and cannot be eliminated by taking necessary steps at the planning, production or inspection stage;
(iii) wear-out or fatigue, caused by the aging of the item.

Since the item is likely to fail at any time, it is quite customary to assume that the life of the item is a random variable with a distribution function $F(t)$ which is the probability that the item fails before time t. Many of the questions raised above can be answered if we know $F(t)$. For example, the average life could be defined as the mean of the distribution $F(t)$ while the probability of failure-free operation between $[t_0, t_0 + t]$, given that the item was 'alive' or working at time t_0 is given by

$$\frac{F(t_0 + t) - F(t_0)}{1 - F(t_0)}$$

Another very important function associated with the failure distribution $F(t)$ is the hazard rate denoted by $\mu(t)$. Consider the probability of failure-free operation within the interval $[t, t + h]$, where h is infinitesimal. If $f(t)$ denotes the probability density function (p.d.f.) corresponding to $F(t)$, then the hazard rate or the 'instantaneous failure rate' is given by

$$\mu(t) = \lim_{h \to 0} \frac{F(t + h) - F(t)}{h\{1 - F(t)\}} = \frac{f(t)}{1 - F(t)}$$

The function $\mu(t)$ is also known as the force of mortality in actuarial and life contingency problems.

Let $R(t) = 1 - F(t)$ denote the probability of failure-free operation until time t, or survival until time t. Then it is quite evident that the stochastic behaviour of the failure time can be studied through either of the four functions $f(t)$, $F(t)$, $R(t)$ or $\mu(t)$. Note that

$$R(t) = 1 - F(t), f(t) = dF(t)/dt \quad \text{and} \quad \mu(t) = -d/dt \log \{1 - F(t)\}$$

From simple integration it follows that

$$F(t) = 1 - \exp\left\{-\int_0^t \mu(w) \, dw\right\} \quad \text{and} \quad f(t) = \mu(t) \exp\left\{-\int_0^t \mu(w) \, dw\right\}$$

We list below several common forms of $f(t)$ that are generally assumed in life testing experiments and reliability problems. The simplest form is the exponential distribution with

$$f(t) = \frac{1}{\sigma} \exp\left(-\frac{t}{\sigma}\right), \, t \geqslant 0, \, \sigma > 0$$

for which $F(t) = 1 - \exp(-t/\sigma)$, $R(t) = \exp(-t/\sigma)$, $\mu(t) = 1/\sigma$.

Davis (1952) examined different types of data and the exponential distribution appears to fit most of the situations quite well. Indeed Epstein (1958) remarks that the exponential distribution plays as important a role in life testing experiments as the part played by the normal distribution in agricultural experiments on effects of different treatments on the yield. We will consider the exponential distribution in detail in Chapter 1. Different sampling schemes such as censoring and truncation will be discussed in this and subsequent chapters.

In Chapter 2 we will introduce more complex forms. First, we consider the gamma distribution with p.d.f.

$$f(t \mid \sigma, p) = \frac{1}{\Gamma(p)\sigma^p} t^{p-1} \exp\left(-\frac{t}{\sigma}\right), \quad t \geqslant 0, \, p > 0, \, \sigma > 0$$

where $\Gamma(p)$ is the well-known gamma function. Note that for $p = 1$, the gamma distribution reduces to the exponential distribution. For the gamma distribution,

$$F(t) = \frac{1}{\Gamma(p)} \int_0^t \frac{t^{p-1}}{\sigma^p} \exp\left(-\frac{t}{\sigma}\right) dt$$

and there is no explicit formula (closed expression) for $F(t)$. It may be noted that $F(t)$ is the well known incomplete gamma integral and has been studied extensively. Similarly, there is no simple formula for the instantaneous failure rate $\mu(t)$ although it can be shown that for $p > 1$, $\mu(t)$ is an increasing function of t. This implies the 'aging effect', i.e., the failure or the hazard rate increases with the time (age) t. This property makes the gamma distribution applicable to many life testing experiments in which the "aging effect" is expected. [See Birnbaum and Saunders (1958), Gupta (1960), Greenwood and Durand (1960), Kendall and Stuart (1972)].

Next, we will consider the Weibull distribution for which the p.d.f.

$$f(t \mid \sigma, k) = (k/\sigma)t^{k-1} \exp(-t^k/\sigma), \ t \geqslant 0, \ k > 0, \ \sigma > 0$$

Again note that for $k = 1$, the Weibull distribution reduces to the exponential distribution. We also note that the Weibull distribution arises in a natural way from the exponential distribution if we assume that the kth power of the failure time has exponential distribution. For the Weibull distribution we have

$$F(t) = 1 - \exp\left(-\frac{t^k}{\sigma}\right), \quad \mu(t) = \frac{kt^{k-1}}{\sigma}$$

Thus, for the Weibull distribution the hazard rate is an increasing function of time and increases as a power of t for $k > 1$; [See Weibull (1939, 1951), Mendenhall and Lehman (1960), Mennon (1963), Cohen (1965), Harter and Moore (1965), Mann (1968), Lawless (1972)].

In Chapter 3 we consider the situation where the failure time follows either the normal or the log-normal distribution. Here the densities are, for the

(i) normal distribution

$$f(t \mid \mu, \sigma) = \frac{1}{\sqrt{2\pi}\sigma} \exp\left\{-\frac{(t-\mu)^2}{2\sigma^2}\right\}, \quad -\infty < t < \infty,$$

$$-\infty < \mu < \infty, \ \sigma > 0$$

(ii) log-normal distribution

$$f(t \mid \mu, \sigma) = \frac{1}{t\sigma\sqrt{2\pi}} \exp\left\{\frac{-(\log t - \mu)^2}{2\sigma^2}\right\}, \quad t > 0, \ \sigma > 0,$$

$$-\infty < \mu < \infty$$

The expressions for $\mu(t)$ or $F(t)$ for these distributions cannot be obtained in a closed form. For the normal we can show that $\mu(t)$ is an increasing function of time and, therefore, the model implies the aging-effect. For log-normal distribution, however, $\mu(t)$ is a decreasing function of time t. This implies that the hazard rate decreases with time and therefore log-normal distribution indicates a reversal of the aging effect. We refer to papers by Gupta (1952), Aitchison and Brown (1957), Plackett (1959), Bazovsky (1961), Larson (1969) where the use of normal and/or the related distributions in the above context is extensively discussed.

In Chapter 4 we consider mixtures and compound distributions. In Chapter 5 we will discuss the problem of testing of hypotheses and construction of confidence intervals for various life testing models. In Chapter 6 we will consider a Bayesian approach to life testing experiments and reliability when the failure time distributions are exponential, Weibull and normal. [See Bhattacharya (1967), Draper and Guttman (1972), Bogdanoff and Pierce (1973), Sinha and Guttman (1976), which consider some situations in which the Bayesian approach is suitable and appropriate.] Finally, in Chapter 7 we will briefly discuss the reliability of series and parallel systems.

The book concludes with two appendices. Appendix 1 outlines the basic distribution theory and some important results used. Appendix 2 covers the basic theory of estimation and tests of hypotheses.

CHAPTER 1

Exponential Failure Model

Any inference about the average life is based on the data that are assumed to be drawn from a universe or population specified by a distribution function (d.f.), $F(x)$. Before one approaches the general problem of life testing, one has to make some assumptions about the underlying $F(x)$ or its corresponding p.d.f., $f(x)$. In life testing research the simplest and the most widely exploited model is the one-parameter exponential distribution with p.d.f.

$$f(x \mid \sigma) = (1/\sigma) \exp(-x/\sigma), \; x \geqslant 0, \; \sigma > 0 \tag{1}$$

Here σ is the average or the mean life of the item and it also acts as a scale parameter.

Exponential distribution plays an important part in life testing problems as mentioned in the 'Introduction'. For a situation where the failure rate appears to be more or less constant, the exponential distribution would be an adequate choice but not all items satisfy the condition that 'it does not age'. There are several situations where the failure rate may be increasing or decreasing and Weibull, gamma or log-normal would be a more realistic choice. Given the data, perhaps the best one can do is to apply some transformation which will support the assumption that the transformed observations are exponentially distributed (Draper and Guttman, 1965), or check the assumption of exponentiality by some appropriate statistical test (Epstein, 1960). Exponential distribution also occurs in several other contexts, such as the waiting time problems. Maguire, Pearson and Wynn (1952) studied mine accidents and showed that time intervals between accidents follow exponential distribution. Let X be the life of an item under test. The exponential distribution may be easily derived by using the relationship

$$f(x) = \mu(x) \exp\left\{-\int_0^x \mu(w) \, dw\right\}$$

A constant failure-rate λ yields

$$f(x \mid \lambda) = \lambda \exp(-\lambda x), \; x, \lambda > 0$$

There are, however, some other elementary considerations which lead to an exponential distribution. These considerations may be formally stated as assumptions:

(1) The failure of the item in a given interval of time $[t_0, t_1]$ depends only on $(t_1 - t_0)$, the length of the time interval and not on t_0, the position of the time interval.

(2) Suppose the probability that the item will fail in an infinitesimal time interval $[t, t + h]$ is proportional to λ except for infinitesimals of higher order, then the probability of failure in the above time interval is $\lambda h + 0(h)$.

(3) The probability of failure at $t = 0$, i.e. the moment the test started, is zero.

Let

$$R(t) = P(x \geqslant t)$$

= the probability that the item survives for at least time t

(assumption 2)

In view of the assumptions (1) and (2) we write

$$R(t + h) = R(t)(1 - \lambda h) + 0(h)$$

where λ is a constant.

$$[R(t + h) - R(t)]/h = -\lambda R(t) + 0(h)/h$$

Taking limits as $h \to 0$ we get a simple differential equation

$$dR(t)/dt = -\lambda R(t)$$

or

$$d/dt \log R(t) = -\lambda$$

the solution of which is

$$R(t) = A \exp(-\lambda t)$$

where A is an arbitrary constant.

From the assumption (3),

$$R(0) = 1 = A$$

Hence

$$R(t) = \exp(-\lambda t) \tag{2}$$

$$F(t) = 1 - R(t) = 1 - \exp(-\lambda t)$$

and

$$f(t) = \lambda \exp(-\lambda t), \, t, \lambda > 0$$

Some Properties of Exponential Distribution

The exponential distribution has several interesting properties. We mention a few below:

(i) The distribution is 'forgetful' or 'has no memory'. What it means, however, is that if a unit has survived t hours, then the probability of its surviving an additional h hours is exactly the same as the probability of surviving h hours of a new item.

Consider the one-parameter exponential density with mean life σ, viz.,

$$f(x \mid \sigma) = (1/\sigma) \exp(-x/\sigma), \, x, \sigma > 0.$$

$$P(X \geqslant t + h \mid x \geqslant t) = \frac{\int_{t+h}^{\infty} \frac{1}{\sigma} \exp\left(-\frac{x}{\sigma}\right) dx}{\int_{t}^{\infty} \frac{1}{\sigma} \exp\left(-\frac{x}{\sigma}\right) dx} = \exp\left(-\frac{h}{\sigma}\right) = P(X \geqslant h)$$

(ii) Suppose n items are under test *with replacements* and the failure time distribution is exponential with mean life σ; then the between failure times are independent and identically distributed as exponential with mean life σ/n. (We will frequently use the terms 'with or without replacement' to refer to respective situations where the items that fail 'are or are not replaced by similar new items'.)

Let $X_{(1)} < X_{(2)} \ldots < X_{(n)}$ be the ordered failure times of n items under test and let the failure times be exponentially distributed with mean life σ. The test starts at time $X_{(0)} = 0$ and the system operates till $X_{(1)} = x_{(1)}$ when the first failure occurs. Suppose the failed item is immediately replaced by a new item and the system operates till the second failure occurs at time $X_{(2)} = x_{(2)}$ and the item failed is immediately replaced. The process continues till all items fail and let

$$W_1 = X_{(1)}, \ W_2 = X_{(2)} - X_{(1)}, \ \ldots, \ W_n = X_{(n)} - X_{(n-1)}$$

Now W_1 is distributed as the first order statistic $X_{(1)}$ in a sample of n from the exponential p.d.f. (1). Hence the p.d.f. of W_1 may be written down as

$$g(w_1 \mid \sigma) = (n/\sigma) \exp(-nw_1/\sigma), \ 0 < w_1 < \infty \tag{3}$$

which implies that W_1 is exponentially distributed with mean life (σ/n).

Since the exponential distribution is 'memory-less' and the items that fail are immediately replaced, $W_2, W_3, \ldots, W_n$ are independent and identically distributed (i.i.d.) as W_1. (See Appendix 1 for discussion on the distribution of order statistics.)

(iii) If n items are put to test *without replacement* and $(X_{(1)}, X_{(2)}, \ldots, X_{(n)})$ are ordered failure times from an exponential population with mean life σ, then $(Z_1, Z_2, \ldots, Z_n)$ are i.i.d. as

$$g(z \mid \sigma) = (1/\sigma) \exp(-z/\sigma), \ z, \sigma > 0$$

where $Z_i = (n - i + 1)\{X_{(i)} - X_{(i-1)}\}, \ i = 1, 2, \ldots, n; \ X_{(0)} = 0.$

Note the difference between (i) and (ii). Under the 'with replacement' plan, the number of items exposed at any time is n and the total time on the test till the kth failure time $x_{(k)}$ is $n\{x_{(1)} + x_{(2)} - x_{(1)} + x_{(3)} - x_{(2)} + \ldots + x_{(k)} - x_{(k-1)}\} = nx_{(k)}$ whereas, when the failed times are not replaced, the number of items exposed at $x_{(0)}$ is n, at $x_{(1)}$ is $(n-1)$, at $x_{(k)}$ it is $(n-k)$, etc., and the total time on the test up to $x_{(1)}$ is $nx_{(1)}$, up to $x_{(2)}$ it is $(n-1)\{x_{(2)} - x_{(1)}\}$, up to $x_{(k)}$ it is $(n-k+1)\{x_{(k)} - x_{(k-1)}\}$.

The joint distribution of $\{X_{(1)}, X_{(2)}, \ldots, X_{(n)}\}$ is given by

$$g(x_{(1)}, x_{(2)}, \ldots, x_{(n)} \mid \sigma)$$

$$= \frac{n!}{\sigma^n} \exp\left\{\frac{\sum_{1}^{n} x_{(i)}}{\sigma}\right\}, \quad 0 < x_{(1)} < x_{(2)} < \ldots < x_{(n)} < \infty \tag{4}$$

Let

$$z_1 = nx_{(1)}$$

$$z_2 = (n - 1)\{x_{(2)} - x_{(1)}\}$$

$$z_3 = (n - 2)\{x_{(3)} - x_{(2)}\}$$

$$\vdots$$

$$z_k = (n - k + 1)\{x_{(k)} - x_{(k-1)}\}, \quad k = 1, 2, \ldots, n; \ x_{(0)} = 0 \qquad (5)$$

$$\frac{\partial(z_1, z_2, \ldots, z_n)}{\partial(x_{(1)}, x_{(2)}, \ldots, x_{(n)})} = n!, \ \sum_1^n z_i = \sum_1^n x_{(i)}, \ z_i > 0, \ i = 1, 2, \ldots, n$$

The Jacobian of transformation is $1/n!$. Substituting in (4), we obtain the joint distribution of $(z_1, z_2, \ldots, z_n)$

$$f(z_1, z_2, \ldots, z_n \mid \sigma) = \frac{1}{\sigma^n} \exp\left(-\sum_1^n z_i\right)$$

$$= \prod_{i=1}^n \frac{1}{\sigma} \exp\left(-z_i/\sigma\right)$$

(iv) Suppose n items are under test and the failure time distribution is exponential with mean life σ. The replacement of items that fail by new items makes the test a Poisson process with intensity $\lambda = n/\sigma$.

Let $N(t)$ be the number of failures that occurred during the interval $[0, t]$. Suppose the probability of failure during $[t, t + h]$ is proportional to λ and the interval is so small that the probability of two or more failures in that interval may be neglected, then the process $\{N(t) \mid t \geqslant 0\}$ is known as the Poisson process with intensity λ.

From (3) it follows that $(nW_1)/\sigma$ is distributed as a gamma variate with parameter 1. We will often express it by the symbol $(nW_i/\sigma) \sim \gamma(1)$. To prove (iv) we will use the following property of gamma distribution: If $(X_1, X_2, \ldots, X_k)$ are i.i.d. as gamma with parameters $n_1, n_2, \ldots, n_k$, then $\sum_1^k X_i$ is distributed as a gamma with a parameter $\sum_1^k n_i$. (See Appendix 1 for proof.) From the above property it follows that $\left(n \sum_1^k W_i \middle/ \sigma\right) \sim \gamma(k)$. Going back to property (ii)

$$P[N(t) = k \mid t]$$

$$= P[W_1 + W_2 + \ldots + W_k \leqslant t] - P[W_1 + W_2 + \ldots + W_{k+1} \leqslant t]$$

$$= P[W_1 + W_2 + \ldots + W_{k+1} \geqslant t] - P[W_1 + W_2 + \ldots + W_k \geqslant t]$$

$$= \frac{1}{\Gamma(k+1)} \int_t^\infty \left\{\exp\left(-\frac{ny}{\sigma}\right)\right\}\left(\frac{ny}{\sigma}\right)^k d\left(\frac{ny}{\sigma}\right)$$

$$- \frac{1}{\Gamma(k)} \int_t^\infty \left\{\exp\left(-\frac{ny}{\sigma}\right)\right\}\left(\frac{ny}{\sigma}\right)^{k-1} d\left(\frac{ny}{\sigma}\right)$$

$$= \exp\left\{-\left(\frac{nt}{\sigma}\right)\right\}\left[\sum_{r=0}^k \left(\frac{nt}{\sigma}\right)^r \middle/ r! - \sum_{r=0}^{k-1} \left(\frac{nt}{\sigma}\right)^r \middle/ r!\right]$$

$$= \frac{(\lambda t)^k}{k!} \exp\left(-\lambda t\right), \ \lambda = \frac{n}{\sigma}$$

Estimation of Mean Life with Complete Samples

Suppose n items are subjected to test and the test is terminated after all the items have failed. Let $(X_1, X_2, \ldots, X_n)$ be the random failure times and suppose the failure times are exponentially distributed with p.d.f. $f(x \mid \sigma) = (1/\sigma) \exp(-x/\sigma)$, $x, \sigma > 0$. Given the sample $(X_1, X_2, \ldots, X_n)$ we wish to estimate the mean life σ.

The likelihood function is given by

$$L(x_1, x_2, \ldots, x_n \mid \sigma) \equiv L = \frac{1}{\sigma^n} \exp\left(-\sum_1^n x_{i/\sigma}\right)$$

$$\frac{\partial}{\partial \sigma} \log L = -\frac{n}{\sigma} \frac{\sum_1^n x_i}{\sigma^2}$$

The maximum likelihood estimator (MLE) of σ is the solution of the equation

$$\frac{\partial}{\partial \sigma} \log L = 0$$

Thus, $\hat{\sigma} = \bar{x}$ is the MLE of σ. Also $E(\hat{\sigma}) = E(\bar{X}) = \sigma$ and Var $(\bar{X}) = \sigma^2/n$. Note that $\bar{x}$ is *unbiased* for σ. Consider the distribution of $\bar{X}$. Since $(X/\sigma) \sim \gamma(1)$, by a property of gamma distribution mentioned earlier, it follows that

$$\sum_1^n X_i/\sigma = \frac{n\bar{X}}{\sigma} \sim \gamma(n)$$

and we write

$$g(\bar{x} \mid \sigma) = \frac{1}{\Gamma(n)} \left(\frac{n\bar{x}}{\sigma}\right)^{n-1} \left(\frac{n}{\sigma}\right) \exp\left(-\frac{n\bar{x}}{\sigma}\right)$$

$$= \frac{\left(\frac{n}{\sigma}\right)^n}{\Gamma(n)} \bar{x}^{n-1} \exp\left(-\frac{n\bar{x}}{\sigma}\right) \quad \bar{x} > 0 \tag{6}$$

We now consider the minimum variance unbiased estimator of σ and indicate the proof of the fact that $\hat{\sigma} = \bar{x}$ is also the uniformly minimum variance unbiased estimator (UMVUE) of σ. In general, the search for UMVUE is limited to the functions of the complete sufficient statistic when it exists. The sufficiency of any statistic can be checked by the factorizibility criterion, but the completeness is much more difficult to establish. In the single parameter exponential distribution, the sufficiency of $\bar{x}$ is easy to check. Consider the ratio

$$\frac{L(x_1, x_2, \ldots, x_n \mid \sigma)}{g(\bar{x} \mid \sigma)}$$

which in this case reduces to

$$\frac{\Gamma(n)}{n\left(\sum_1^n x_i\right)^n} \quad \text{for} \quad x_i > 0, \ i = 1, 2, \ldots, n$$

This ratio is independent of σ and is a function of observations alone. Hence $\bar{x}$ is sufficient. For the proof of completeness of $\bar{x}$, we refer to Lehman and Scheffé (1955). Thus $\bar{x}$ is UMVUE of σ.

In this particular case, both methods of estimation lead to the same estimator. But we will see in the next section that this is not so when we are estimating the reliability $R(t) = \exp(-t/\sigma)$, which is the probability of survival for at least time t.

Reliability Estimation

We use the term 'reliable' in various contexts in everyday life, such as a reliable friend, a reliable service station, reliable news, etc. As an abstract concept it means something or someone we may depend upon or count on. In life testing research we are more concerned with a quantitative measure of the reliability of an item or a device we are interested in.

The reliability of a unit (or a system) is defined as the probability that it will perform satisfactorily at least for a specified period of time without a major breakdown. If X is the lifetime of the unit, the reliability of the unit at time t is given by

$$R(t) = P(X \geq t) = 1 - F(t)$$

where F is the d.f. of the failure time X.

Suppose a manufacturer wants to promote a new brand of light bulb in the market. A random sample of 10 bulbs is put to test and their failure times recorded. Suppose the bulbs failed after 125, 189, 210, 351, 465, 580, 630, 760, 810 and 870 hours. Looking at the data, a prospective buyer asks, "If I buy a new bulb of this brand, what is the probability that it will survive at least 600 hours?"

If we do not assume any particular distribution of the failure times, this probability may be estimated by the ratio

$$[R(t)]_{t=600} = \frac{\text{number of bulbs surviving} \geq 600 \text{ hours}}{\text{number of bulbs initially exposed to test}} = \frac{4}{10} = 0.4$$

If we assume that the failure time distribution is exponential with mean life σ, by definition

$$R(t) = \int_t^\infty \frac{1}{\sigma} \exp(-x/\sigma)\, dx = \exp(-t/\sigma)$$

Since σ is unknown, we have to estimate $R(t)$. First consider the MLE. We have already seen that the MLE of σ is given by $\hat{\sigma} = \bar{x}$. Now MLE has an important property that if $\hat{\theta}$ is the MLE of θ, then $\tau(\hat{\theta})$ is the MLE of $\tau(\theta)$ provided $\tau(\theta)$ is a fairly well behaved function of θ. In particular say, $\tau(\theta)$ is a monotone differentiable function of θ (see Zehna, 1966). Thus we have $\hat{R}(t) = \exp(-t/\bar{x})$. From the failure times of 10 bulbs, we obtain $\bar{x} = 499$, and hence the MLE of reliability at time $t = 600$ hours is given by

$$\hat{R}(600) = \exp\left(-\frac{600}{499}\right) = \exp(-1.2024) = 0.3006$$

Although $\hat{R}(t)$ is easy to compute, like most maximum likelihood estimators, $\hat{R}(t)$ is biased and we will see that actual derivation of $E\{\hat{R}(t)\}$ and Var $\{\hat{R}(t)\}$ is difficult, and uses a complicated special function. Using the distribution of $\bar{x}$ in (6),

$$E\{\hat{R}(t)\} = \frac{\left(\frac{n}{\sigma}\right)^n}{\Gamma(n)} \int_0^\infty \bar{x}^{n-1} \exp\left\{-\left(\frac{t}{\bar{x}} + \frac{n\bar{x}}{\sigma}\right)\right\} d\bar{x}$$

$$= \frac{1}{\Gamma(n)} \int_0^\infty y^{n-1} \exp\left\{-\left(y + \frac{nt}{\sigma y}\right)\right\} dy, \; y = \frac{n\bar{x}}{\sigma} \qquad (7)$$

To evaluate the integral we recall a result due to Watson (1952), viz.,

$$\int_0^\infty \bar{x}^r \exp\left\{-\left(ax + \frac{b}{x}\right)\right\} dx = 2\left(\frac{a}{b}\right)^{(r-1)/2} K_{r-1}(2\sqrt{ab})$$

where $K_r(\quad)$ is the modified Bessel function of the second kind of order r. Also $K_{-r}(\quad) = K_r(\quad)$ for $r = 0, 1, 2, \ldots$. Substituting $-r = n - 1$, $a = 1$, $b = nt/\sigma$ above, we obtain

$$E\{\hat{R}(t)\} = \frac{2}{(n-1)!}\left(\frac{nt}{\sigma}\right)^{n/2} K_n\left\{2\left(\sqrt{\frac{nt}{\sigma}}\right)\right\}$$

Similarly,

$$E\{\hat{R}(t)\}^2 = \frac{2}{(n-1)!}\left(\frac{2nt}{\sigma}\right)^{n/2} K_n\left\{2\left(\sqrt{\frac{2nt}{\sigma}}\right)\right\}$$

and hence the Var $\{\hat{R}(t)\}$ can be obtained.

The uniformly minimum variance unbiased estimator of reliability for a number of distributions has been worked out by several authors. Pugh (1963) obtained UMVUE of $R(t)$ for the one-parameter and Laurent (1963) and Tate (1959) for the two-parameter exponential model with complete samples. Basu (1964) considered the Weibull and Gamma distributions (with known shape parameter) as well as the exponential distribution when testing is terminated after the first r failures. Sinha (1972) studied the behaviour of the UMVUE of $R(t)$ for the one-parameter exponential model, when a spurious observation may be present and suggested an alternative estimator.

The general method of finding the UMVUE is to look for any unbiased statistic $T(x_1, x_2, \ldots, x_n)$ and a complete sufficient statistic $\hat{\theta}$ if one exists. Then the UMVUE is given by $E[T(x_1, x_2, \ldots, x_n) \mid \hat{\theta}]$.

Given the random failure times $(X_1, X_2, \ldots, X_n)$ from the exponential p.d.f. $f(x \mid \sigma) = 1/\sigma \exp(-x/\sigma)$, $x, \sigma > 0$, we want to obtain $\tilde{R}(t)$, the UMVUE of $P(x \geqslant t)$. Consider a function $T(x_1, \ldots, x_n)$ such that $T(x_1) = 1$ if $x_1 \geqslant t$ and $= 0$ otherwise. Thus T is a function of x_1 alone and we denote it by $T(x_1)$.

$$E\{T(x_1)\} = 1 \cdot P(X_1 \geqslant t) + 0 \cdot P(X_1 < t) = \int_t^\infty f(x_1 \mid \sigma) \, dx_1 = R(t \mid \sigma)$$

Thus $T(x_1)$ is an unbiased estimator of $R(t)$. We have noted earlier that $\bar{x}$ is a complete sufficient statistic for σ. Hence the UMVUE

$$\tilde{R}(t) = E\{T(x_1) \mid \bar{x}\} = \int_t^\infty f(x_1 \mid \sigma, \bar{x}) \, dx \tag{8}$$

We now derive the conditional distribution of X_1, given $\bar{x}$. Split the random sample $(X_1, X_2, \ldots, X_n)$ into two independent components $X = X_1$, a sample of size 1 and $(X_2, X_3, \ldots, X_n)$, a sample of size $(n-1)$; $\dfrac{X_1}{\sigma} \sim \gamma(1)$ and $\dfrac{(n-1)\bar{Y}}{\sigma} \sim \gamma(n-1)$ where

$$\bar{Y} = \frac{\sum\limits_{j=2}^n X_j}{n-1} \quad \text{and} \quad (n-1)\bar{Y} + X_1 = \sum_1^n X_i = n\bar{X} \tag{9}$$

Joint distribution of X_1 and $\bar{Y}$ is given by

$$f(x_1, \bar{y} \mid \sigma) = \frac{(n-1)^{n-1}}{\sigma^n \Gamma(n-1)} \, \bar{y}^{n-2} \exp\left[-\frac{1}{\sigma}\{(n-1)\bar{y} + x_1\}\right]$$

since x_1 and $\bar{y}$ are independent.

By virtue of the relationship in (9), we obtain the joint distribution of X_1 and $\bar{X}$, by using a simple transformation

$$f(x_1, \bar{x} \mid \sigma) = \frac{n(n-1)^{n-2}}{\sigma^n \Gamma(n-1)} \left[\frac{n\bar{x}}{n-1} - \frac{x_1}{n-1}\right]^{n-2} \exp\left(-\frac{n\bar{x}}{\sigma}\right), \; 0 < x_1 < n\bar{x}$$

Going back to (6),

$$g(\bar{x} \mid \sigma) = \left(\frac{n}{\sigma}\right)^n \frac{1}{\Gamma(n)} \, \bar{x}^{n-1} \exp\left(-\frac{n\bar{x}}{\sigma}\right), \; \bar{x} > 0$$

$$f(x_1 \mid \bar{x}, \sigma) = \frac{f(x_1, \bar{x} \mid \sigma)}{g(\bar{x} \mid \sigma)}$$

$$= \frac{n-1}{n\bar{x}} \left(1 - \frac{x_1}{n\bar{x}}\right)^{n-2}, \; 0 < x_1 < n\bar{x}$$

From (8),

$$\tilde{R}(t) = \frac{n-1}{n\bar{x}} \int_t^{n\bar{x}} \left(1 - \frac{x}{n\bar{x}}\right)^{n-2} dx$$

$$= \left(1 - \frac{t}{n\bar{x}}\right)^{n-1}, \; t < n\bar{x}$$

and $\qquad\qquad = 0, \; t \geq n\bar{x}$ \hfill (Basu, 1964)

Since $\tilde{R}(t)$ is the UMVUE of $R(t)$, $E(\tilde{R}(t)) = \exp(-t/\sigma)$.

Now

$$\frac{S}{\sigma} \sim \gamma(n), \text{ where } S = \sum_{i=1}^n X_i$$

Hence,

$$E\{\tilde{R}^2(t)\} = \frac{1}{\sigma^n \Gamma(n)} \int_t^\infty s^{n-1} \left(1 - \frac{t}{s}\right)^{2n-2} \exp\left(-\frac{s}{\sigma}\right) ds$$

$$= \frac{(t/\sigma)^n \exp(-t/\sigma)}{\Gamma(n)} I$$

where

$$I = \int_0^\infty \frac{u^{2n-2} \exp(-ut/\sigma)}{(u+1)^{n-1}} du$$

$$= \sum_{i=0}^{2n-2} a_i \int_0^\infty \frac{(u+1)^i \exp(-ut/\sigma)}{(u+1)^{n-1}} du$$

$$= \sum_{i=0}^{n-1} a_i \int_0^\infty \frac{\exp(-ut/\sigma)}{(u+1)^{n-1-i}} du + \sum_{i=n}^{2n-2} a_i \int_0^\infty (u+1)^{i-n+1} \exp(-ut/\sigma) du$$

and

$$a_i = (-1)^i \binom{2n-2}{i}.$$

Using the following results (Erdelyi, 1954):

$$- E_i(-x) = \int_x^\infty \frac{\exp(-u)}{u} du$$

and

$$\int_0^\infty \frac{\exp(-up) du}{(u+a)^n} = \sum_{m=1}^{n-1} \frac{(m-1)! (-p)^{n-m-1}}{(n-1)! a^m} - \frac{(-p)^{n-1}}{(n-1)!} \exp(ap) \cdot E_i(-ap)$$

$$\text{Var}\{\tilde{R}(t)\} = \frac{(t/\sigma)^n \exp(-t/\sigma)}{\Gamma(n)} \left[\sum_{i=0}^{n-3} a_i \left\{ \sum_{m=1}^{n-i-2} \frac{(m-1)!}{(n-i-2)!} \left(-\frac{t}{\sigma}\right)^{n-m-i-2} \right. \right.$$
$$\left. - \frac{(-t/\sigma)^{n-i-2}}{(n-i-2)!} E_i\left(-\frac{t}{\sigma}\right) \exp\left(\frac{t}{\sigma}\right) \right\} - a_{n-2} E_i\left(-\frac{t}{\sigma}\right) \exp\left(\frac{t}{\sigma}\right)$$
$$\left. + \frac{a_{n-1}}{t/\sigma} + \sum_{i=n}^{2n-2} \frac{(i-n+1)! \, a_i \, i^{-n+1} (t/\sigma)^r}{(t/\sigma)^{i-n+2}} \sum_{r=0}^{} \frac{(t/\sigma)^r}{r!} \right] - \exp\left(-\frac{2t}{\sigma}\right)$$

A program has been developed which will compute $\text{Var}\{\tilde{R}(t)\}$ for any combination of n and (t/σ) by using a 10-point generalized Gauss-Laguerre quadratic formula to evaluate I. Table (1.1) computes $\text{Var}\{\tilde{R}(t)\}$ for $(t/\sigma) = (1/6), (1/6), (18/6)$ and $n = 2, (2), 20$ (Sinha, 1972).

For the data on the new brand of bulbs mentioned earlier

$$\tilde{R}(600) = \left(1 - \frac{600}{4990}\right)^9 = 0.3158 \text{ while } \hat{R}(600) = 0.3066$$

We see that the actual values of $\hat{R}(t)$ and $\tilde{R}(t)$ differ only slightly. The difference between the two estimates will be still smaller for large n since the exponential function $\exp(-a)$ approximates well the series $(1 - a/n)^n$ for large values of n.

EXAMPLE 1.1 Twenty electronic tubes were put to test and the test continued till all of them failed. The failure times (in hours) were recorded as follows:

$$9.9, \quad 35.6, \quad 57.9, \quad 94.6, \quad 141.4,$$
$$154.4, \quad 163.3, \quad 226.7, \quad 244.3, \quad 337.2,$$
$$391.8, \quad 417.2. \quad 444.6, \quad 461.2, \quad 497.1,$$
$$582.6, \quad 606.8, \quad 616.3, \quad 672.0, \quad 784.7.$$

Assuming the failure times to be distributed exponentially with mean life σ, compute
  (i) the MLE of $R(t)$,
 (ii) the UMVUE of $R(t)$ at $t = 1000$ hours.
Estimate the standard error of the estimator (ii).
We have

$$n = 20 \sum_1^{20} x_i = 6939.6, \; \bar{x} = 346.98$$

At $t = 1000$ hours, we have

$$\text{MLE } \hat{R}(t) = \exp(-t/\bar{x})$$
$$= \exp(-2.882) = 0.056$$
$$\text{UMVUE } \tilde{R}(t) = \left(1 - \frac{t}{\sum_1^{20} x_i}\right)^{n-1} = \left(\frac{5939.6}{6939.6}\right)^{19}$$
$$= 0.052$$

As expected, there is little difference between $\hat{R}(t)$ and $\tilde{R}(t)$ since n is fairly large. Using Table 1.1, the estimated standard error of $\hat{R}(t) = 0.036$.

Table 1.1*   Var (R_t^*) for $n = 2(2)20$ and $t/\sigma = (1/6)(1/6)(18/6)$

t/σ \ n	2	4	6	8	10	12	14	16	18	20
1/6	0.0270	0.0081	0.0045	0.0031	0.0024	0.0019	0.0016	0.0014	0.0012	0.0011
2/6	0.0566	0.0204	0.0121	0.0085	0.0066	0.0053	0.0045	0.0039	0.0034	0.0031
3/6	0.0756	0.0300	0.0184	0.0132	0.0103	0.0084	0.0071	0.0062	0.0054	0.0049
4/6	0.0848	0.0357	0.0224	0.0163	0.0128	0.0105	0.0089	0.0077	0.0068	0.0061
5/6	0.0868	0.0380	0.0242	0.0177	0.0140	0.0115	0.0098	0.0085	0.0076	0.0068
6/6	0.0841	0.0377	0.0243	0.0179	0.0142	0.0117	0.0100	0.0087	0.0077	0.0069
7/6	0.0785	0.0357	0.0232	0.0172	0.0136	0.0113	0.0096	0.0084	0.0075	0.0067
8/6	0.0714	0.0327	0.0213	0.0158	0.0126	0.0105	0.0090	0.0078	0.0069	0.0062
9/6	0.0637	0.0292	0.0191	0.0142	0.0113	0.0094	0.0081	0.0071	0.0063	0.0056
10/6	0.0560	0.0257	0.0168	0.0125	0.0110	0.0083	0.0071	0.0062	0.0055	0.0050
11/6	0.0487	0.0222	0.0146	0.0180	0.0086	0.0072	0.0062	0.0054	0.0048	0.0043
12/6	0.0420	0.0190	0.0124	0.0092	0.0074	0.0061	0.0052	0.0046	0.0041	0.0037
13/6	0.0359	0.0162	0.0105	0.0078	0.0062	0.0052	0.0044	0.0039	0.0034	0.0031
14/6	0.0305	0.0136	0.0088	0.0065	0.0052	0.0043	0.0037	0.0032	0.0029	0.0026
15/6	0.0259	0.0114	0.0073	0.0054	0.0043	0.0036	0.0030	0.0027	0.0024	0.0021
16/6	0.0218	0.0095	0.0061	0.0045	0.0035	0.0029	0.0025	0.0022	0.0019	0.0017
17/6	0.0183	0.0078	0.0050	0.0037	0.0029	0.0024	0.0020	0.0018	0.0016	0.0014
18/6	0.0154	0.0065	0.0041	0.0030	0.0024	0.0019	0.0017	0.0014	0.0013	0.0011

*Reproduced with the kind permission of the Editor of *Operations Research*.

Comparing $\hat{R}(t)$ and $\tilde{R}(t)$ we note that $\hat{R}(t)$ is biased while $\tilde{R}(t)$ is not. The expression for variance for both of these estimators is quite complicated. A general method of comparison between a biased estimator and an unbiased one is to consider the mean squared error (MSE) of the two estimators. Zacks and Even (1966) have compared the $\hat{R}(t)$ and $\tilde{R}(t)$ for sample size $n = 4$ and $n = 8$, and it is shown that for $t/\sigma < 0.25$, $\tilde{R}(t)$, i.e., UMVUE has smaller MSE than that of $\hat{R}(t)$, the MLE. However, for t/σ between 0.5 to 3.5, $\hat{R}(t)$ has smaller MSE than $\tilde{R}(t)$. Thus, we cannot prefer one estimate to the other for all values of t/σ. Note that σ is in fact unknown, and the above information can only be used on the basis of some vague information that may be available about t/σ.

The advantages and disadvantages of both these methods of estimation have been discussed in several papers and in many situations. The general conclusion appears to be that one cannot say that one method is preferable to another, under all circumstances. We have, therefore, discussed both the methods—and the decision as to which one to use in a given situation is left with the reader, since it would depend on several factors relevant to the situation. We would also like to point out that in many cases both methods lead to the same estimator. One disadvantage of UMVUE is that such an estimator can take values which are inappropriate. Thus, $\tilde{R}(t) = 0$ for $t > n\bar{x}$ while the model assigns a positive value to $R(t \mid \sigma)$. In some analysis of variance situations UMVUE leads to a negative estimate of the variance which is essentially a positive quantity.

We now consider a slight generalization of the exponential distribution which 'naturally' arises as follows:

In the exponential distribution we have assumed that the probability of failure at the starting time is zero. In many situations one can reasonably assume that the probability of failure up to a certain time say μ is zero and, thereafter, the failure time follows the exponential distribution. This implies that p.d.f. of the failure time X is given by

$$f(x \mid \mu, \sigma) = \frac{1}{\sigma} \exp\left(-\frac{x - \mu}{\sigma}\right), \ x \geqslant \mu, \ \sigma > 0$$

and $\qquad\qquad\qquad = 0$ elsewhere $\qquad\qquad\qquad\qquad\qquad$ (10)

If μ is known, say $\mu = \mu_0$ (which we may take to be zero without any loss of generality), then we obtain the single parameter exponential distribution with mean σ. However, in most cases μ is unknown and one of the objectives of life testing experiments would be to estimate μ. It is quite evident that the knowledge about μ may be used to write guarantees/warranties about the failure-free time of the item under consideration.

As usual, let $(X_1, X_2, \ldots, X_n)$ be a sample of life times of n items and let $\{X_{(1)}, X_{(2)}, \ldots, X_{(n)}\}$ be the corresponding order statistics, i.e., the lifetimes arranged according to increasing order of magnitude. We first consider the maximum likelihood estimators of μ and σ. The likelihood of the sample is given by

$$L(x_1, x_2, \ldots, x_n \mid \mu, \sigma) = \frac{1}{\sigma^n} \exp \left\{ - \frac{\sum\limits_1^n (x_i - \mu)}{\sigma} \right\}, \quad x_i \geqslant \mu, \quad i = 1, 2, \ldots, n$$

and $\qquad\qquad\qquad\qquad = 0$ elsewhere

Consider σ fixed, say $= \sigma_0$ and consider the maximization of $L(x_1, x_2, \ldots$ $\ldots, x_n \mid \mu, \sigma_0)$ as μ varies over the permissible range $\mu \leqslant x_i, i = 1, 2, \ldots, n$. It is quite clear that $\underset{\mu}{\text{Max}}\, L(x_1, x_2, \ldots, x_n \mid \mu, \sigma_0)$ is attained at $\hat{\mu}$ which minimizes $\sum\limits_1^n (x_i - \mu)$. Thus $\hat{\mu}$ is the maximum possible value of μ subject to $\mu \leqslant x_i, i = 1, 2, \ldots, n$ or $\hat{\mu} = x_{(1)} = \text{Min}\,(x_1, x_2, \ldots, x_n)$. We note that $\hat{\mu}$ is independent of σ_0 and, therefore, of σ.

Now, consider

$$L(x_1, x_2, \ldots, x_n \mid \hat{\mu}, \sigma) = \frac{1}{\sigma^n} \exp \left\{ - \frac{\sum\limits_1^n (x_i - x_{(1)})}{\sigma} \right\}, \quad x_i \geqslant x_{(1)}$$

and $\qquad\qquad\qquad\qquad = 0$ elsewhere.

$\underset{\sigma}{\text{Max}}\, L(x_1, x_2, \ldots, x_n \mid \sigma, x_{(1)})$ is attained at $\hat{\sigma} = \frac{1}{n} \sum\limits_{i=1}^n (x_i - x_{(1)})$.

Thus, the MLE of μ and σ are

$$\hat{\mu} = x_{(1)} \qquad \text{and} \qquad \hat{\sigma} = \bar{x} - x_{(1)}$$

We now study the distribution theory of $\hat{\mu}$ and $\hat{\sigma}$. Consider the random variables

$$Z_i = (n - i + 1)\{X_{(i)} - X_{(i-1)}\}, \quad i = 1, 2, \ldots, n \text{ with } X_{(0)} = \mu$$

$$\hat{\mu} = x_{(1)} = \frac{z_1}{n} + \mu$$

$$\hat{\sigma} = \frac{1}{n} \sum_{i=2}^n (x_i - x_{(1)}) = \frac{\sum\limits_{i=2}^n z_i}{n}$$

It immediately follows, therefore, that $\hat{\mu}$ and $\hat{\sigma}$ are independent and $\text{Cov}\,(\hat{\mu}, \hat{\sigma}) = 0$.

We have seen earlier, that $Z_1, Z_2, \ldots, Z_n$ are i.i.d. exponential random variables with mean σ. Hence

$$E(\hat{\mu}) = \mu + \frac{\sigma}{n}, \quad \text{Var}\,(\hat{\mu}) = \frac{\sigma^2}{n^2}$$

$$E(\hat{\sigma}) = \frac{n-1}{n}\,\sigma, \quad \text{Var}\,(\hat{\sigma}) = \frac{(n-1)\sigma^2}{n^2}$$

Next we consider the minimum variance unbiased estimation (MVUE) approach. We note that $\hat{\mu}$ and $\hat{\sigma}$ are biased, but the bias may be easily corrected.

$$\tilde{\sigma} = \frac{n\hat{\sigma}}{n-1} = \frac{n(\bar{x} - x_{(1)})}{n-1}$$

is unbiased for σ and

$$\text{Var}(\tilde{\sigma}) = \left(\frac{n}{n-1}\right)^2, \ \text{Var}(\tilde{\sigma}) = \frac{\sigma^2}{n-1};$$

similarly

$$\tilde{\mu} = \hat{\mu} - \frac{\tilde{\sigma}}{n} = \hat{\mu} - \frac{\hat{\sigma}}{n-1} = \frac{nx_{(1)} - \bar{x}}{n-1}$$

is unbiased for μ and

$$\text{Var}(\tilde{\sigma}) = \text{Var}(\hat{\mu}) + \text{Var}\left(\frac{\hat{\sigma}}{n-1}\right) = \frac{\sigma^2}{n(n-1)};$$

$$\text{Cov}(\tilde{\mu}, \tilde{\sigma}) = \text{Cov}\left(\hat{\mu} - \frac{\hat{\sigma}}{n-1}, \frac{n\hat{\sigma}}{n-1}\right)$$

$$= -\frac{n}{(n-1)^2}\text{Var}(\hat{\sigma})$$

$$= -\frac{\sigma^2}{n(n-1)}$$

$\tilde{\mu}$ and $\tilde{\sigma}$ are functions of $\left\{x_{(1)}, \sum_2^n (x_{(i)} - x_{(1)})\right\}$ which can be shown to be jointly complete and sufficient for (μ, σ). This implies that $\tilde{\mu}$ and $\tilde{\sigma}$ are UMVUE of μ and σ respectively. We indicate below the proof of sufficiency and for the property of completeness we refer to Lehman and Scheffé (1955). The proof of completeness uses some rather deep results from real and complex variable theory.

Let $U = X_{(1)}, V = \sum_{i=2}^n (X_{(i)} - X_{(1)})$. As indicated earlier, U and V are independent random variables and the p.d.f.s of U and V are given by

$$g_1(u \mid \mu, \sigma) = \frac{n}{\sigma}\exp\left(-\frac{n(u-\mu)}{\sigma}\right), u \geq \mu$$

and $$g_2(v \mid \sigma) = \frac{1}{\Gamma(n-1)}\frac{v^{n-2}}{\sigma^{n-1}}\exp\left(-\frac{v}{\sigma}\right), v \geq 0, \sigma > 0$$

The joint density of U and V is given by $g_1(u \mid \mu, \sigma) \cdot g_2(v \mid \sigma)$.

The ratio $\dfrac{L(x_1, x_2, \ldots, x_n \mid \mu, \sigma)}{g_1(u \mid \mu, \sigma)g_2(v/\sigma)} = \dfrac{\Gamma(n-1)}{nv^{n-2}}$ is independent of μ and σ and is a function of observations. Thus $\{x_{(1)}, \sum_{i=2}^n (x_{(i)} - x_{(1)})\}$ are jointly sufficient for (μ, σ).

Suppose we are interested in estimating the expected life when the failure times are distributed as two-parameter exponential (10). The expected life is given by $E(X) = \mu + \sigma$, the MLE of which is given by $\hat{\mu} + \hat{\sigma} = x_{(1)} + \bar{x} - x_{(1)} = \bar{x}$, while the UMVUE of the expected life is given by

$$\tilde{\mu} + \tilde{\sigma} = \hat{\mu} - \frac{\hat{\sigma}}{n-1} + \frac{n\hat{\sigma}}{n-1} = \hat{\mu} + \hat{\sigma} = \bar{x}$$

In this case the MLE and UMVUE are the same.

Consider the estimation of the reliability function

$$R(t \mid \mu, \sigma) = 1 \quad \text{if } t \leqslant \mu$$

and

$$= \exp\left\{-\left(\frac{t-\mu}{\sigma}\right)\right\}, \quad t > \mu$$

Here the maximum likelihood approach immediately leads to

$$\hat{R}(t) = R(t \mid \hat{\mu}, \hat{\sigma})$$

$$= 1 \quad \text{if } t \leqslant x_{(1)}$$

and

$$= \exp\left(-\frac{t - x_{(1)}}{\bar{x} - x_{(1)}}\right), \quad t > x_{(1)}$$

On the other hand, to obtain the UMVUE of $R(t \mid \mu, \sigma)$ is a lot more difficult than that in the case of a single parameter exponential distribution (with $\mu = 0$). We, therefore, just quote the result and refer to Laurent (1963) for details of derivation:

The UMVUE of $R(t \mid \mu, \sigma)$ is given by

$$\tilde{R}(t) = 1 \quad \text{if } t \leqslant x_{(1)}$$

$$= \left(1 - \frac{1}{n}\right)\left[1 - \frac{t - x_{(1)}}{v}\right]^{n-2}, \quad x_{(1)} < t \leqslant v + x_{(1)}$$

and

$$= 0, \qquad\qquad t > v + x_{(1)}$$

where

$$v = \sum_{i=2}^{n} (x_{(i)} - x_{(1)})$$

In rather rare circumstances, for the two-parameter exponential case one may know σ (but not μ), which we may take without any loss of generality to be unity. The failure time distribution is given by

$$f(x \mid \mu, 1) = \exp\{-(x - \mu)\}, x \geqslant \mu$$

The problem of obtaining the MLE and UMVUE of μ and the reliability function

$$R(t \mid \mu) = \exp\{-(t - \mu)\}, t \geqslant \mu$$

is left as an exercise for the readers.

Estimation with Censored Samples

So far, we have assumed that we have a complete sample available where the failure times of all the n items are recorded. There are several situations, however, where this is neither possible nor desirable. Note that life-testing experiments are usually destructive in that the items are destroyed at the end of the experiment and cannot be used again. This limits the number of items we can test. We may put n items on test and terminate the experiment when a pre-assigned number of items, say r ($<n$) have failed. The samples obtained from such an experiment are

called 'failure-censored' samples. Failure-censored sampling is almost mandatory in dealing with high cost sophisticated items such as colour television tubes.

Another factor that affects the life-testing experiment is the amount of time required to obtain the complete sample. To limit this factor, we may put n items to test and terminate the experiment at a pre-assigned time t_0. The samples obtained from such an experiment are called 'time-censored' samples. Time-censored sampling is almost essential in dealing with life-testing experiments in which the cost of experiments increases heavily with time.

In the failure-censored case data consist of the life times of the r items that failed (say $x_{(1)} < x_{(2)} < \ldots < x_{(r)}$) and the fact that $(n - r)$ items have survived beyond $x_{(r)}$. In the time-censored case data consist of the life times of items that failed before the time t_0, say $x_{(1)}x_{(2)} < \ldots < x_{(m)}$, assuming that m items failed before t_0 and the fact that $(n - m)$ items have survived beyond t_0. In the time-censored case t_0, the time of termination is fixed while m, the number of items that failed before t_0, is a random variable. In the failure-censored case the situation is reverse in that r, the number of items that failed, is fixed, while $x_{(r)}$, the time at which the experiment is terminated, is a random variable. One may, of course, consider a combination of both time and failure censoring where one terminates the experiment when either a pre-assigned number r of items have failed or the experiment has run up to time t_0, whichever comes first. This approach is useful when the items involved are high cost ones and the time element in obtaining the observations is also important. For example, consider a new type of rotary engine. This is a high cost item and would consume fuel and require trained technicians as observers. On the other hand, consider polyethylene bags which are being tested for their breaking strength. In this case one would hardly recommend either time or failure-censored sampling.

We will now consider the estimation of parameters and reliability functions for one and two-parameter exponential distributions under different types of censoring commonly used in life-testing experiments.

Failure-censored Samples

Let us first consider the case where the items that failed are not replaced.

We consider the single parameter exponential distribution first. As stated earlier, the data consist of failure times $x_{(1)} < x_{(2)} < \ldots < x_{(r)}$ of r items that failed and the fact that $(n - r)$ items survived until $x_{(r)}$. The likelihood of the sample is given by

$$L\{x_{(1)}, x_{(2)}, \ldots, x_{(r)} \mid \sigma\}$$

$$= \frac{n(n - 1)(n - 2) \ldots (n - r + 1)}{\sigma^r} \exp \left\{ - \frac{\sum_1^r x_{(i)} + (n - r)x_{(r)}}{\sigma} \right\} \quad (11)$$

This can be obtained from the following considerations.

For the complete sample case, the joint p.d.f. of $X_{(1)}, X_{(2)}, \ldots, X_{(n)}$ is given by

$$g\{x_{(1)}, x_{(2)}, \ldots, x_{(n)}\} = \frac{n!}{\sigma^n} \exp\left(\frac{\sum_1^n x_{(i)}}{\sigma}\right), \quad 0 < x_{(1)} < x_{(2)} \ldots < x_{(n)} \quad (12)$$

In the failure-censored case we terminate the experiment at $x_{(r)}$ and so we integrate out (12) with respect to $x_{(r+1)}, x_{(r+2)}, \ldots, x_{(n)}$ over the region $x_{(r)} < x_{(r+1)} < x_{(r+2)} < \ldots < x_{(n)}$ which immediately leads to (11).

The MLE of σ is easily shown to be $\hat{\sigma}_r = \left(\sum_1^r x_{(i)} + (n-r)x_{(r)}\right)\bigg/r.$

Using the transformation $z_i = (n - i + 1)\{x_{(i)} - x_{(i-1)}\}$, $i = 1, 2, \ldots, r$ with $x_{(0)} = 0$, $\hat{\sigma}_r = \left(\sum_1^r z_i\right)\bigg/r$. We have shown earlier that $(Z_1, Z_2, \ldots, Z_r)$ are i.i.d. as $f(z/\sigma) = \frac{1}{\sigma} \exp(-z/\sigma)$, $z, \sigma > 0$.

Hence $\quad \frac{r\hat{\sigma}_r}{\sigma} \sim \gamma(r), \quad E\left(\frac{r\hat{\sigma}_r}{\sigma}\right) = \mathrm{Var}\left(\frac{r\hat{\sigma}_r}{\sigma}\right) = r.$ Thus $\quad E(\hat{\sigma}_r) = \sigma$ and $\mathrm{Var}(\hat{\sigma}_r) = \frac{\sigma^2}{r}$ and the p.d.f. of $\hat{\sigma}_r$ is given by

$$g(\hat{\sigma}_r \mid \sigma) = \frac{1}{\Gamma(r)} \left(\frac{r}{\sigma}\right)^r (\hat{\sigma}_r)^{r-1} \exp\left(-\frac{r\hat{\sigma}_r}{\sigma}\right), \quad 0 < \hat{\sigma}_r < \infty \quad (13)$$

Consider the ratio

$$\frac{L\{x_{(1)}, x_{(2)}, \ldots, x_{(r)} \mid \sigma\}}{g(\hat{\sigma}_r \mid \sigma)} = \frac{\Gamma(r)n(n-1)(n-2)\ldots(n-r+1)}{(r)^r \hat{\sigma}_r^{r-1}}$$

which is independent of the unknown parameter σ and thus $\hat{\sigma}_r$ is sufficient for σ. Also, it is known to be complete (Tukey, 1949; Smith, 1957). Since $E(\hat{\sigma}_r) = \sigma$, it follows, therefore, that $\hat{\sigma}_r$ is also the UMVUE of σ. The MLE of the reliability function $R(t \mid \sigma)$ is given by

$$\hat{R}(t) = \exp(-t/\hat{\sigma}_r)$$

Following the same line of derivation as in the complete sample case, we obtain

$$g(x_i \mid \hat{\sigma}_r) = \frac{\Gamma(r)}{\Gamma(r-1)r\hat{\sigma}_r} \left(1 - \frac{x_i}{r\hat{\sigma}_r}\right)^{r-2}, \quad 0 < x_i < r\hat{\sigma}_r,$$

where x_i is the failure time of any of the r units that failed. The UMVUE of reliability, $\tilde{R}(t)$ is given by

$$R(t) = P(X_i > t \mid \hat{\sigma}_r)$$

$$= \frac{\Gamma(r)}{\Gamma(r-1)r\hat{\sigma}_r} \int_t^{r\hat{\sigma}_r} \left(1 - \frac{x_1}{r\hat{\sigma}_r}\right)^{r-2} dx_i$$

$$= \left(1 - \frac{t}{r\hat{\sigma}_r}\right)^{r-1}, \quad t < r\hat{\sigma}_r$$

and

$$= 0, \quad t \geqslant r\hat{\sigma}_r$$

Next, consider the case where the items that fail are immediately replaced and the test terminates after the rth failure. Here the likelihood function is given by

$$L\{x_{(1)}, x_{(2)}, \ldots, x_{(r)} \mid \sigma\} = \frac{n^r}{\sigma^r}\left[\exp\left\{-\frac{x_{(1)}}{\sigma}\right\}\right]\left[\exp\left\{-\frac{x_{(2)} - x_{(1)}}{\sigma}\right\}\right]\cdots$$

$$\left[\exp\left\{-\frac{x_{(r)} - x_{(r-1)}}{\sigma}\right\}\right]\left[\int_{x_{(r)}}^{\infty}\frac{1}{\sigma}\exp\left(-\frac{x}{\sigma}\right)dx\right]^{n-1}$$

$$= \left(\frac{n}{\sigma}\right)^r \exp\left\{-\frac{nx_{(r)}}{\sigma}\right\}, \quad 0 < x_{(1)} < x_{(2)}\ldots < x_{(r)} < \infty$$

The MLE $\sigma_r^* = \dfrac{nx_{(r)}}{r}$. Also

$$\frac{r\sigma_r^*}{\sigma} = \frac{n}{\sigma}[x_{(1)} + \{x_{(2)} - x_{(1)}\} + \{x_{(3)} - x_{(2)}\} + \{x_{(r)} - x_{(r-1)}\}] \sim \gamma(r)$$

Thus, in the experiment of either type (with or without replacement) the MLE of σ is unbiased, has the same variance and has the same distribution. $\dfrac{L\{x_{(1)}, x_{(2)}, \ldots, x_{(r)} \mid \sigma\}}{g(\sigma_r^* \mid \sigma)} = \dfrac{n\Gamma(r)}{rx_{(r)}^{r-1}}$ is independent of σ and thus σ_r^* is sufficient. It is also known to be complete. Hence σ_r^* is the MLE as well as UMVUE of σ. The MLE of reliability function $R(t \mid \sigma)$ is given by $R^*(t) = \exp(-t/\sigma_r^*)$. Since σ_r^* is complete and sufficient, if we can find a function $h\{\sigma_r^* \mid t\}$ such that $E\{h(\sigma_r^* \mid t)\} = \exp(-t/\sigma)$, then the UMVUE of reliability is given by $h(\sigma_r^* \mid t)$.

Consider the function

$$h(\sigma_r^* \mid t) = \left(1 - \frac{t}{r\sigma_r^*}\right)^{r-1}, \quad t < r\sigma_r^*$$

and

$$= 0, \quad t \geqslant r\sigma_r^*$$

From (13),

$$g(\sigma_r^* \mid \sigma) = \frac{1}{\Gamma(r)}\left(\frac{r}{\sigma}\right)^r (\sigma_r^*)^{r-1}\exp\left(-\frac{r\sigma_r^*}{\sigma}\right), \quad 0 < \sigma_r^* < \infty$$

$$E\{h(\sigma_r^* \mid t)\} = \frac{1}{\Gamma(r)}\left(\frac{r}{\sigma}\right)^r\int_{t/r}^{\infty}\left(1 - \frac{t}{r\sigma_r^*}\right)^{r-1}(\sigma_r^*)^{r-1}\exp\left(-\frac{r\sigma_r^*}{\sigma}\right)d\sigma_r^*$$

$$= \frac{r}{\Gamma(r)\sigma^r}\int_{t/r}^{\infty}(r\sigma_r^* - t)^{r-1}\exp\left(-\frac{r\sigma_r^*}{\sigma}\right)d\sigma_r^*$$

$$= \frac{1}{\Gamma(r)\sigma^r}\int_0^{\infty} y^{r-1}\exp\left\{-\frac{1}{\sigma}(y + t)\right\}dy$$

$$= \exp(-t/\sigma)$$

This shows that $h(\sigma_r^* \mid t)$ is the UMVUE of $R(t \mid \sigma)$.

EXAMPLE 1.2 Sixty items were placed on test and the test was terminated after the first 10 items failed. The failure times (in hours) were recorded as follows: 85, 151, 280, 376, 492, 520, 623, 715, 820 and 914.

Assuming the failure time distribution is exponential

$$f(x \mid \sigma) = \frac{1}{\sigma} \exp(-x/\sigma), \; x, \; \sigma > 0$$

estimate σ and the reliability function at $t = 600$ hours if the failed items are:

(i) not replaced,
(ii) replaced.

We have $n = 60$, $r = 10$. Under the case (i),

$$\hat{\sigma}_r = \frac{\sum_1^{10} x_{(i)} + 50x_{(10)}}{10} \approx 5068 \text{ hours}$$

$$\hat{R}(600) = \exp\left(-\frac{600}{5068}\right) = \exp(-0.118) = 0.8887$$

$$\tilde{R}(600) = \left(1 - \frac{600}{50680}\right)^9 = (0.9882)^9 = 0.8987$$

Under (ii),

$$\sigma_r^* = \frac{60 \times 914}{10} = 5484$$

$$\hat{R}(600) = \exp\left(-\frac{600}{5484}\right) = \exp(-0.109) = 0.8967$$

$$\tilde{R}(600) = \left(1 - \frac{600}{54840}\right)^9 = (0.9891)^9 = 0.9060$$

Samples	Estimates			Data:
	$\hat{\sigma}_r$ (or σ_r^*)	$\hat{R}(t)$	$\tilde{R}(t)$	$n = 60$
Without replacement	5068	0.8887	0.8987	$r = 10$
With replacement	5484	0.8967	0.9060	$t = 600$

Since

$$\hat{\sigma}_r = \frac{\sum_1^r x_{(i)} + (n-r)x_{(r)}}{r} = \frac{\sum_1^r x_{(i)} - rx_{(r)}}{r} + \sigma_r^* \tag{14}$$

we have $\hat{\sigma}_r < \sigma_r^*$ and the estimate of $R(t \mid \sigma)$ based on $\hat{\sigma}_r$ is less than the one based on σ_r^*.

Sampling with or without replacement, has its own merits and limitations. The replacement method is expensive since it presupposes that sufficient spares are available and also requires trained technicians who can do the necessary replacement as soon as an item fails. Without replacement, the test is easy to perform when neither special assistance nor costly replacement parts are needed. Under the same expected failure rates, the replacement plan has a slight edge over the non-replacement counterpart in that it requires a smaller sample to start with, which follows immediately from (14). But the saving seldom compensates the cost involved.

Consider the data in Example 1.2, without replacement case. We have $n = 60, r = 10, \hat{\sigma}_r = 5068$. For with replacement case

$$ n = \frac{r\sigma_r^*}{x_{(r)}} = \frac{10 \times 5068}{914} \simeq 55, \text{ a saving of 8.33 per cent} $$

If instead of a single parameter exponential we assume that failure times are distributed as two-parameter model (10), then the likelihood of the sample is

$$ L\{x_{(1)}, x_{(2)}, \ldots, x_{(r)} \mid \mu, \sigma\} $$

$$ = \frac{n(n-1)(n-2)\ldots(n-r+1)}{\sigma^r} \exp\left\{\frac{\sum\limits_{i=1}^{r}(x_{(i)} - \mu) + (n-r)(x_{(r)} - \mu)}{\sigma}\right\} $$

$$ \tag{15} $$

where $\mu \leqslant x_{(1)} < x_{(2)} \ldots < x_{(r)} < \infty$. To obtain the MLE of μ and σ we first fix σ and maximize (15) with respect to μ as μ varies over the range $\mu \leqslant x_{(1)}$. We must select μ so as to minimize $\sum\limits_{i=1}^{r}\{x_{(i)} - \mu\} + (n-r)\{x_{(r)} - \mu\}$ $= \sum\limits_{i=1}^{r} x_{(i)} + (n-r)x_{(r)} - n\mu$. This minimum is attained at $\hat{\mu} = x_{(1)}$ and does not depend upon the fixed value of σ. Now consider $L\{x_{(1)}, x_{(2)}, \ldots, x_{(r)} \mid \hat{\mu}, \sigma\}$. It can be seen that this is maximized at $\hat{\sigma}_r = \dfrac{y_r}{r}$, where

$y_r = \sum\limits_{i=1}^{r} x_{(i)} + (n-r)x_{(r)} - nx_{(1)}$. The p.d.f. of $\hat{\mu} = x_{(1)}$ is easily obtained as $g_1\{x_{(1)} \mid \mu, \sigma\} = \dfrac{n}{\sigma} \exp\left\{-\dfrac{n(x_{(1)} - \mu)}{\sigma}\right\}$, $x_{(1)} \geqslant \mu$. To obtain the p.d.f. of Y_r, we consider the transformation

$$ Z_i = (n - i + 1)\{X_{(i)} - X_{(i-1)}\}, \quad i = 1, 2, \ldots, r $$

where $X_{(0)} = \mu$. $Z_1, Z_2, \ldots, Z_r$ are i.i.d. random variables distributed as

$f(z \mid \sigma) = \dfrac{1}{\sigma} \exp(-z/\sigma), z, \sigma > 0$. Hence $\dfrac{Y_r}{\sigma} = \dfrac{\sum\limits_{i=2}^{r} Z_i}{\sigma} \sim \gamma(r-1)$ with p.d.f.

$g_2(y_r \mid \sigma) = \dfrac{1}{\Gamma(r-1)} \left(\dfrac{y_r}{\sigma}\right) \dfrac{1}{\sigma} \exp\left(\dfrac{-y_r}{\sigma}\right)$, $0 < y_r < \infty$. Note that $X_{(1)}$ and Y_r are independent and so are $\hat{\mu}$ and $\hat{\sigma}_r$.

$$ E(\hat{\sigma}_r) = \frac{\sigma}{r} E\left(\frac{Y_r}{\sigma}\right) = \frac{r-1}{r} \sigma \text{ and Var}(\hat{\sigma}_r) = \frac{\sigma^2}{r^2} \text{Var}\left(\frac{Y_r}{r}\right) = \frac{(r-1)\sigma^2}{r^2} $$

Also $X_{(1)} = \dfrac{Z_1}{n} + \mu$. Hence $E(\hat{\mu}) = \mu + \dfrac{\sigma}{n}$, $\text{Var}(\hat{\mu}) = \dfrac{\sigma^2}{n^2}$. Thus $\hat{\mu} = x_{(1)}$ and

$\hat{\sigma}_r = y_r/r$ are biased estimators of μ and σ respectively. The sufficiency of $\{x_{(1)}, y_r/r\}$ may be checked by considering the ratio

$$\frac{L\{x_{(1)}, x_{(2)}, \ldots, x_{(r)} \mid \mu, \sigma\}}{g_1\{x_{(1)} \mid \mu, \sigma\}g_2(y_r \mid \sigma)} = \frac{n(n-1)(n-2)\ldots(n-r+1)\Gamma(r-1)}{y_r^{r-2}}$$

which is independent of μ and σ. $\{x_{(1)}, y_r/r\}$ is further known to be complete (Turkey, 1949; Smith, 1957). Hence the estimators $\tilde{\sigma}_r = \dfrac{y_r}{r-1}$, $\tilde{\mu} = x_{(1)}$ $-\dfrac{y_r}{n(r-1)}$ are UMVUE of σ and μ respectively and $\text{Var}(\tilde{\sigma}_r) = \dfrac{\sigma^2}{r-1}$, $\text{Var}(\tilde{\mu}) = \dfrac{r\sigma^2}{n^2(r-1)}$.

The above results agree with those for complete samples $r = n$. Let us compare the estimators obtained from censored samples with those obtained from complete samples. For the single parameter exponential case, these are $\hat{\sigma}_r = \dfrac{1}{n}\left\{\sum_1^r x_{(i)} + (n-r)x_{(r)}\right\}$ and $\hat{\sigma}_n = \dfrac{1}{n}\sum_1^n x_{(i)}$ respectively. We have seen that both are unbiased for σ but $\text{Var}(\hat{\sigma}_n) = (\sigma^2/n) < \text{Var}(\hat{\sigma}_r) = (\sigma^2/r)$. Thus $\hat{\sigma}_n$ has a smaller variance than $\hat{\sigma}_r$ and therefore $\hat{\sigma}_r$ would have lesser precision than $\hat{\sigma}_n$. This is to be expected since the failure-censored sample data consist of $\{x_{(1)}, x_{(2)}, \ldots x_{(r)}\}$ and $(n-r)$ survivals beyond $x_{(r)}$ while the complete sample data consist of $\{x_{(1)}, x_{(2)}, \ldots x_{(r)}, x_{(r+1)}, \ldots, x_{(n)}\}$, the failure times of all the items. Thus there is loss of information due to censoring the last $(n-r)$ observations $\{x_{(r+1)}, x_{(r+2)}, \ldots x_{(n)}\}$. The larger the value of r, the smaller would be the loss of information and the greater would be the precision of the estimator. The particular choice of r in a given life testing experiment would depend on the cost of items and the degree of precision we would like to have.

EXAMPLE 1.3   Suppose 30 items are under test and the test is terminated after the first 15 items failed. The first failure occurs 650 hours after the test begins and the observed total life time between the first and the fifteenth failure is 14,000 hours. Assuming the failure time distribution is a two-parameter exponential, obtain the MLE and UMVUE of μ and σ.

Let $x_{(1)}, x_{(2)}, \ldots x_{(15)}$ be the failure times of the first 15 items. We have

$x_{(1)} = 650$, $r = 15$, $n = 30$.

$$14,000 = 29\{x_{(2)} - x_{(1)}\} + 28\{x_{(3)} - x_{(2)}\} + \ldots + 16\{x_{(15)} - x_{(14)}\}$$

$$= x_{(2)} + x_{(3)} + x_{(4)} + \ldots + x_{(14)} + 16x_{(15)} - 29x_{(1)}$$

$$= \sum_{i=1}^{15} x_{(i)} + 15x_{(15)} - 30x_{(1)}$$

$$= y_{15}$$

The MLE's are

$$\hat{\mu} = x_{(1)} = 650 \text{ hours}$$

$$\hat{\sigma}_r = \frac{y_r}{r} = \frac{14,000}{15} = 933 \text{ hours}$$

The UMVUE's are

$$\tilde{\mu} = x_{(1)} - \frac{y_r}{n(r-1)} = 650 - \frac{14,000}{30 \times 14} = 617 \text{ hours}$$

$$\tilde{\sigma}_r = \frac{y_r}{r-1} = \frac{14,000}{14} = 1000 \text{ hours}$$

Based on the first r ordered failure times, the MLE of reliability is given by $\hat{R}(t \mid \mu, \sigma) = \exp\left\{-\frac{1}{\hat{\sigma}_r}(t - \hat{\mu})\right\}$ and the UMVUE of $R(t \mid \mu, \sigma)$ is given by

$$\tilde{R}(t \mid \mu, \sigma) = \frac{n-1}{n}\left\{1 - \frac{t - x_{(1)}}{y_r}\right\}^{r-2}, \quad x_{(1)} < t < x_{(1)} + y_r$$

$$= 1, \quad x_{(1)} > t$$

and

$$= 0, \quad t > x_{(1)} + y_r$$

(Laurent, 1963)

EXAMPLE 1.4   Given the data in Example 1.3, obtain the MLE and UMVUE of the reliability at $t = 1000$ hours. We have $\hat{\sigma}_r = 933$, $\hat{\mu} = 650$. Hence

$$\hat{R}(1000 \mid \mu, \sigma) = \exp\left\{-\frac{1000 - 650}{933}\right\} = 0.6873$$

and

$$\tilde{R}(1000 \mid \mu, \sigma) = \frac{29}{30}\left\{1 - \frac{1000 - 650}{14,000}\right\}^{13} = 0.6956$$

Time-censored Samples

The time-censored samples arise when we terminate the life-testing experiment at a pre-assigned time t_0. As mentioned earlier, here the number of items that failed before time t_0 is a random variable which we denote by M. Let $p(t_0)$ be the probability of failure before time t_0. Then M has a binomial distribution

$$P(M = m) = \binom{n}{m}p^m q^{n-m}, \quad m = 0, 1, 2, \ldots, n$$

where $p \equiv p(t_0) = 1 - \exp(-t_0/\sigma)$ and $q = 1 - p$. Suppose the items that failed are *not-replaced*. The data consist of the life times $x_{(1)} < x_{(2)} < \cdots < x_{(m)}$ of m items that failed before t_0 and $(n - m)$ items that survived beyond t_0. The likelihood of the sample is given by

$$L\{x_{(1)}, x_{(2)}, \ldots, x_{(m)}, m \mid \sigma\} = \exp\left(-\frac{nt_0}{\sigma}\right) \text{ for } m = 0$$

and

$$= \frac{n!}{m!\,(n-m)!}\frac{1}{\sigma^m}\exp\left\{-\frac{\sum_{i=1}^{m} x_{(i)} + (n-m)t_0}{\sigma}\right\}$$

for $m = 1, 2, \ldots, n$.

This follows from the following considerations. The likelihood for $m = 0$ is obvious. For $m > 0$ consider the conditional p.d.f. of the failure time, given that the item has failed before time t_0. This is given by

$$h(x \mid \sigma) = \frac{\frac{1}{\sigma} \exp\left(-\frac{x}{\sigma}\right)}{1 - \exp\left(-t_0/\sigma\right)}, \quad 0 < x \leqslant t_0$$
$$= 0 \quad \text{otherwise}$$

Thus, the joint p.d.f. of $x_{(1)}, x_{(2)}, \ldots, x_{(m)}$ is given by

$$g\{x_{(1)}, x_{(2)}, \ldots, x_{(m)} \mid \sigma\} = \frac{m!}{\sigma^m} \frac{\exp\left\{-\sum\limits_{i=1}^{m} x_{(i)}\big/\sigma\right\}}{\left\{1 - \exp\left(-\dfrac{t_0}{\sigma}\right)\right\}^m}$$

The likelihood of the sample is the joint p.d.f. of $\{x_{(1)}, x_{(2)}, \ldots, x_{(m)}\}$ and m. Hence

$$L\{x_{(1)}, x_{(2)}, \ldots, x_{(m)}, m \mid \sigma\} = g\{x_{(1)}, x_{(2)}, \ldots, x_{(m)} \mid \sigma\}\binom{n}{m}p^m q^{n-m}$$

$$= -\frac{n!}{(n-m)!} \frac{1}{\sigma^m} \exp\left[-\frac{\sum\limits_{i=1}^{m} x_{(i)} + (n-m)t_0}{\sigma}\right]$$

For $m > 0$, $L\{x_{(1)}, x_{(2)}, \ldots, x_{(m)}, m \mid \sigma)$ is maximized for

$$\hat{\sigma} = \frac{\sum\limits_{i=1}^{m} x_{(i)} + (n-m)t_0}{m}$$

However, for $m = 0$, the likelihood $= \exp\left(-nt_0/\sigma\right)$ and is a strictly increasing function of σ. So the MLE should be taken to be $\hat{\sigma} = \infty$ if $m = 0$. But then the random variable $\hat{\sigma}$ is improper in that it assumes the value $+\infty$ with positive probability and, therefore, neither the mean nor the variance would exist. We, however, note that $P(M = 0) = \exp\left(-nt_0/\sigma\right)$ would be quite small for a fairly large value of nt_0/σ. This can be achieved by having n sufficiently large and t_0/σ not too small. Thus, in practice for the time-censored life-testing experiment, we will start with fairly large number of items n and would select termination time t_0 such that t_0 is not too small as compared to σ. Since σ is unknown, one cannot guarantee that $m > 0$ for a given choice of n and t_0. There is always a positive probability, however small, that we obtain $m = 0$. In such a case we follow the recommendation of Bartholomew (1957) that we take $\hat{\sigma} = nt_0$. Thus the MLE σ is given by

$$\left. \begin{aligned} \hat{\sigma} &= \frac{\sum\limits_{1}^{m} x_{(1)} + (n-m)t_0}{m}, \quad m > 0 \\ &= nt_0 \quad\quad \text{for} \quad\quad m = 0 \end{aligned} \right\} \tag{16}$$

The distribution theory of $\hat{\sigma}$ in the time-censored case is quite complicated. This is mainly due to the fact that m is a random variable and the numerator of (16) is a sum of random number of random variables, Bartholomew (1957) has obtained approximations to the bias and variance of $\hat{\sigma}$. The asymptotic (large sample) bias and variance are given by

$$\text{Bias}(\hat{\sigma}) \simeq \frac{tq}{np^2} \text{ and } \text{Var}(\hat{\sigma}) \simeq \frac{\sigma^2}{np}$$

The MLE of the reliability is, of course, given by $\hat{R}(t) = \exp\left(-\frac{t}{\hat{\sigma}}\right)$. Not much is known about the bias and the variance of $\hat{R}(t)$. The case of the UMVUE appears to be much more hopeless and to the best of knowledge of the authors, the problem of obtaining the UMVUE of σ for the time-censored case remains unsolved. This situation illustrates the principle that sometimes very (apparently) minor modifications in the experimental set up can lead to quite difficult problems in estimating the parameters.

Estimator based on n and m (Bartholomew, 1963)
Consider only the number of failures during $[0, t_0]$ and ignore the times when failures occurred. We have the likelihood

$$L = \binom{n}{m} p^m q^{n-m}, \quad m = 0, 1, 2, \ldots, n;$$

$$= \binom{n}{m}\left\{1 - \exp\left(-\frac{t_0}{\sigma}\right)\right\}^m \exp\left\{-(n-m)\frac{t_0}{\sigma}\right\}$$

and

$$\frac{\partial}{\partial \sigma} \log L = \frac{t_0}{\sigma^2}\left[n - \frac{m}{1 - \exp(-t_0/\sigma)}\right] - E\left(\frac{\partial^2}{\partial \sigma^2} \log L\right)$$

$$= \frac{t_0^2 n \exp(-t_0/\sigma)}{\sigma^4\{1 - \exp(-t_0/\sigma)\}} \text{ for large } n$$

Thus, the MLE of σ and the asymptotic variance of the estimator are given by

$$\hat{\sigma}' = \frac{-t_0}{\log(1 - m/n)} \tag{17}$$

and

$$\text{Var}(\hat{\sigma}') = \frac{\sigma^2 p}{nq(\log q)^2} \tag{18}$$

The estimator (17) is recommended for $0.2 < m/n < 0.8$. The bias in this estimator is infinite since the probability that $m = n$ is positive (Gnedenko, Belyayev and Solovyev, 1969).

The limiting relative efficiency of $\hat{\sigma}'$ against $\hat{\sigma}$ in (16)

$$= \lim_{n \to \infty} \frac{\text{Var}(\hat{\sigma})}{\text{Var}(\hat{\sigma}')}$$

$$= \frac{q}{p^2}\{\log(1 - p)\}^2$$

$$= (1 - p)\left(1 + p + \frac{11p^2}{12} + \cdots\right)$$

$$= 1 - \frac{p^2}{12} \text{ if } p \text{ is small}$$

Bartholomew (1963) obtained the exact variance of $\hat{\sigma}$ and $\hat{\sigma}'$ for $n = 10$ and showed that for $p < \frac{1}{2}$, $\hat{\sigma}'$ is as efficient as $\hat{\sigma}$.

Gnedenko, et al (1969) notes that for relatively reliable units with $t_0/\sigma < 0.1$, the conditional p.d.f. of the instant of failure in the interval $(0, t_0)$ is given by

$$\frac{\frac{1}{\sigma} \exp\left(-\frac{t_0}{\sigma}\right)}{1 - \exp\left(-\frac{t_0}{\sigma}\right)} \simeq \frac{1}{t_0}$$

Thus X_i is approximately uniformly distributed over $(0, t_0)$. The mean of each $X_i = t_0/2$ and, therefore, we can take

$$\sum_{i=1}^{m} x_i \simeq \frac{t_0 m}{2}$$

From (16),

$$\hat{\hat{\sigma}} \simeq \frac{t_0 m/2 + (n-m)t_0}{m} = \frac{t_0}{m}\left(n - \frac{m}{2}\right) \tag{19}$$

Note that

$$\hat{\sigma}' = \frac{-t_0}{\log\left(1 - \frac{m}{n}\right)}$$

$$= \frac{t_0}{\frac{m}{n} + \frac{m^2}{2n^2} + \frac{m^3}{3n^3} + \cdots}$$

$$\simeq \frac{nt_0}{m}\left(1 - \frac{m}{2n}\right)$$

$$= \frac{t_0}{m}\left(n - \frac{m}{2}\right)$$

$$= \hat{\hat{\sigma}}$$

For $m/n \leqslant 0.1$, $\hat{\hat{\sigma}}$ is often used as an estimator of σ.

EXAMPLE 1.5   150 items were tested for 800 hours and 10 failures were recorded at the following times (in hours): 50, 60, 80, 150, 180, 190, 250, 320, 400, 480.

Estimate σ by using
  (i) the failure times,
  (ii) the number of failures during the interval $[0, 800]$.
Compare the estimators. We have $n = 150$, $t_0 = 800$, $m = 10$.
  (i) Using (16),

$$\hat{\sigma} = \frac{2160 + 140 \times 800}{10} = 11416$$

(ii) From (17),

$$\hat{\sigma}' = \frac{-800}{\log{(14/15)}} = \frac{800}{0.06899} = 11596$$

$$s_1 = \text{estimated s.d. of } \hat{\sigma} \simeq \frac{\hat{\sigma}}{n\{1 - \exp{(-t/\hat{\sigma})}\}} = 3582$$

and

$$s_2 = \text{estimated s.d. of } \hat{\sigma}' \simeq \frac{\hat{\sigma}'\sqrt{\hat{p}}}{\sqrt{n\hat{q}}(\log{\hat{q}})^2} = 3649$$

On the basis of the estimated standard deviations of the estimators, $\hat{\sigma}$ is a better estimator of σ than $\hat{\sigma}'$. Also, from (19),

$$\hat{\hat{\sigma}} \simeq \frac{800}{10}\left(150 - \frac{10}{2}\right)$$

$$= 11600$$

$$\simeq \hat{\sigma}', \text{ as expected}$$

The problem outlined here is a particular case of a more general problem Bartholomew (1957) considered. Suppose n items of an equipment are installed on different dates, and some of the items are still working while the rest have failed. Given the dates of installations and the lives of the items, we want to estimate the mean life σ. Let t_i be the time elapsed since the ith item has been installed and let x_i be the life of this item. Each item is assumed to have the same exponential life distribution with mean life σ. The MLE of σ is given by

$$\hat{\sigma} = \frac{1}{m}\sum_{i=1}^{n}\{a_i x_i + (1 - a_i)t_i\} \tag{20}$$

where

$$a_i = 1 \quad \text{if } x_i \leqslant t_i$$
$$= 0 \quad \text{if } x_i > t_i$$

and $\sum_{1}^{n} a_i = m$ (Bartlett, 1953a, b; Bartholomew, 1957; Littel, 1952)

and

$$\text{Var}\,(\hat{\sigma}) = \frac{\sigma^2}{\sum\limits_{1}^{n} p_i} \tag{21}$$

where

$$p_i = 1 - \exp{(-t_i/\sigma)}$$

Putting $t_1 = t_2 = \ldots = t_m = t_0$, we obtain the estimator in (16), etc.

EXAMPLE 1.6   Results of a life test on 10 pieces of equipment

Item No *	1	2	3	4	5	6	7	8	9	10
Date of installation	11 June	21 June	22 June	2 July	21 July	31 July	31 July	1 Aug.	2 Aug	10 Aug.
Date of failure	13 June	—	12 Aug.	—	23 Aug.	27 Aug.	14 Aug.	25 Aug.	6 Aug.	—
Life in days (x_i)	2	(119)	51	(77)	33	27	14	24	4	(37)
No. of days to end of failure (t_i)	81	72	70	60	41	31	31	30	29	21

An estimate is required on August 31. (The three lives in the parentheses would not be known.)

$t_i/\hat{\sigma}$	1.84	1.64	1.59	1.36	0.93	
$1 - \exp(-t_i/\hat{\sigma})$	0.84118	0.80602	0.79607	0.74334	0.60624	
$t_i/\hat{\sigma}$	0.71	0.71	0.68	0.66	0.48	Total
$1 - \exp(-t_i/\hat{\sigma})$	0.50589	0.50589	0.49440	0.48263	0.37936	6.16102

From (20) we have

$$\hat{\sigma} = \frac{\text{total of all lives to 31st August, complete and incomplete}}{\text{total number of complete lives}}$$

$$= \tfrac{1}{7}(2 + 51 + 33 + 27 + 14 + 24 + 4 + 72 + 60 + 21)$$

$$= 44 \text{ days}$$

and from (21), the estimated standard deviation

$$= \frac{\hat{\sigma}}{\sum\limits_i \hat{p}_i}$$

$$= \frac{44}{6.16102}$$

$$= 7.14 \text{ days}$$

Based on the failures of all lives,

$$\hat{\sigma} = \tfrac{1}{10}(2 + 119 + 51 + 77 + 33 + 27 + 14 + 24 + 4 + 37)$$

$$= 38.8 \text{ days}$$

Suppose we carry out a time-censored test where the items that fail are immediately *replaced*. We have seen that one of the properties of exponential model [property (iv)] is that under such a plan the number of failures m has a Poisson distribution with parameter $\lambda = (nt_0)/\sigma$. We

*Reproduced with the kind permission of Bartholomew (1957) and the Editor of the *Journal of the American Statistical Association*.

write

$$P(m \mid \sigma) = \frac{\left(\dfrac{nt_0}{\sigma}\right)^m}{m!} \exp\left(-\frac{nt_0}{\sigma}\right), \quad m = 0, 1, 2, \ldots$$

The MLE of σ is given by $\hat{\sigma} = \dfrac{nt_0}{m}$, $m > 0$ and we accept nt_0 as the MLE of σ for $m = 0$ (Bartholomew, 1957).

m is sufficient as well as complete. If there exists a function $\phi(m)$ such that $E\{\phi(m)\} = \sigma$, then $\phi(m)$ is the UMVUE of σ.

$\sum\limits_{m=0}^{\infty} \phi(m)P(m \mid \sigma) = \sigma$ yields

$$\sum_{m=0}^{\infty} \frac{\phi(m)}{m!} \left(\frac{nt_0}{\sigma}\right)^m = \sigma + \sigma \sum_{m=0}^{\infty} \left(\frac{nt_0}{\sigma}\right)^{m+1} \frac{1}{m!}$$

which implies that no such $\phi(m)$ exists.

Let $\psi(m)$ be the UMVUE of $R(t \mid \sigma) = \exp(-t/\sigma)$.

$\sum\limits_{m=0}^{\infty} \psi(m)P(m \mid \sigma) = \exp(-t/\sigma)$ yields

$$\sum_{m=0}^{\infty} \psi(m)\left(\frac{nt_0}{\sigma}\right)^m \frac{1}{m!} = \exp\left\{\frac{nt_0 - t}{\sigma}\right\} = \sum_{m=0}^{\infty} \left(\frac{nt_0 - t}{\sigma}\right)^m \frac{1}{m!}$$

Hence $\psi(m)\left(\dfrac{nt_0}{\sigma}\right)^m = \left(\dfrac{nt_0-t}{\sigma}\right)^m$ and, therefore, $\tilde{R}(t) = \psi(m \mid \sigma) = \left(1 - \dfrac{t}{nt_0}\right)^m$ is the UMVUE of $R(t \mid \sigma)$. Note that for $t > nt_0$, the estimator will assume a negative value for m odd.

$$E\{\tilde{R}(t)\}^2 = \sum_{m=0}^{\infty} \left(\frac{nt_0 - t}{nt_0}\right)^{2m} \left(\frac{nt_0}{\sigma}\right)^m \exp\left(-\frac{nt_0}{\sigma}\right) \frac{1}{m!}$$

$$= \exp\left(-\frac{nt_0}{\sigma}\right) \sum_{m=0}^{\infty} \left[\left(\frac{nt_0 - t}{nt_0}\right)^2 \left(\frac{nt_0}{\sigma}\right)\right]^m \frac{1}{m!}$$

$$= \exp\left[-\frac{nt_0}{\sigma} + \frac{(nt_0 - t)^2}{nt_0\sigma}\right]$$

$$= \exp\left(-\frac{2t}{\sigma} + \frac{t^2}{nt_0\sigma}\right)$$

$$\text{Var}\{\tilde{R}(t)\} = \exp\left(-\frac{2t}{\sigma} + \frac{t^2}{nt_0\sigma}\right) - \exp\left(\frac{-2t}{\sigma}\right)$$

$$= \exp\left(-\frac{2t}{\sigma}\right)\left\{\exp\left(\frac{t^2}{nt_0\sigma}\right) - 1\right\}$$

Joint Censoring

Gnedenko, et al. (1969) considered a joint censoring scheme. Here n items are under test without replacement. The test is terminated at the instant when either the accumulated time on the test equals a pre-assigned number

s_0 or when the rth unit fails and when this happens, the total time on the test is less than s_0 . . . (21).

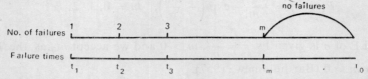

Let the failures occur at times $t_1, t_2, \ldots t_m$. The total time on the test up to t_0 is given by

$$s(t_0) = nt_1 + (n-1)(t_2 - t_1) + (n-2)(t_3 - t_2) + \ldots$$
$$+ (n - m + 1)(t_m - t_{m-1}) + (n - m)(t_0 - t_m)$$
$$= t_1 + t_2 + \ldots + t_m + (n - m)t_0$$

and $s(t_r) =$ total time till the rth failure

$$= nt_1 + (n-1)(t_2 - t_1) + \ldots + (n - r + 1)(t_r - t_{r-1})$$
$$= t_1 + t_2 + \ldots + t_{r-1} + (n - r + 1)t_r$$

The MLE of σ is given by

$$\hat{\sigma} = \frac{s_0}{m} \text{ if } s_0 < s(t_r)$$

$$= \frac{s(t_r)}{r} \text{ if } s_0 > s(t_r)$$

where m is determined by the condition $s(t_0) = s_0$.

Epstein (1961) computed the amount of inspection and expected waiting time for the joint censoring plan (21), for both with and without replacement cases.

EXAMPLE 1.7 Suppose 50 items are under test without replacement. The test stops at the instant when either the accumulated time on the test equals 5000 hours or when the 6th item fails and the total time when this happens < 5000.

Let the first failure occur at $t_1 = 30$.

$$s(t_1) = 50 \times 30 = 1500 < 5000$$

Let the second failure occur at $t_2 = 50$.

$$s(t_2) = 1500 + 49(50 - 30) = 2480 < 5000$$

Let the third failure occur at $t_3 = 70$.

$$s(t_3) = 2480 + 48(70 - 50) = 3440 < 5000$$

Suppose no failure occurs between t_3 and t_0, where

$$5000 = 3440 + 47(t_0 - 70)$$
$$t_0 = 103.2$$

$$\hat{\sigma} = \frac{s_0}{m} = \frac{5000}{3} \simeq 1667 \text{ hours}$$

Estimators Based on Components of Order Statistics

The reader may have noticed that the estimators of μ and σ, which were obtained in the previous section, under a variety of circumstances were all linear functions of the order statistics $x_{(1)}, x_{(2)}, \ldots, x_{(n)}$. This is not too surprising since in life testing experiments, the order statistics arise naturally, when we record the times of first failure $x_{(1)}$, second failure $x_{(2)}$, etc., rather than the failure times of the first (experimental) item x_1, second (experimental) item x_2, etc. There is extensive literature available on the problem of estimation of the parameters by linear combination of order statistics. For the general motivation behind this method of estimation, we refer the reader to discussions contained in Sarhan and Greenberg (1962) and David (1970). As pointed out by Sarndal (1962) and Kulldorff (1962) these estimators also arise in a very natural way in dealing with censored samples and partially grouped samples. The general conclusion appears to be that the loss of efficiency, in terms of increase in variance or mean squared error (over that of the standard estimators) is usually very small. The main advantage appears to be in the increased robustness of the estimator against the presence of outliers and/or spurious observations and in general departures from the assumed model. Kale and Sinha (1971) and Joshi (1972) have studied in great detail the problem of estimating σ for single parameter exponential distribution, when data could have at most one outlier or spurious observation.

The general approach is to consider the estimators based on k specific components of the order statistics. The selection of the specific k components of the order statistics is made on the basis of the smallest possible mean squared error of the estimator. If the condition of unbiasedness is imposed then such a choice is based on the minimum variance. Kulldorff (1962) and Ogawa (1960) studied the asymptotic solutions for one and two-parameter exponential distributions while small sample solutions have been considered by Harter (1961) for $k = 1$ or 2, Sarhan and Greenberg (1962), and Siddiqui (1963) for $k = 2$. For the censored samples we refer the reader to Sarndal (1962), Sarhan and Greenberg (1962), Saleh (1964), and Saleh and Ali (1966). Although the problem of estimation of μ and σ by components of order statistics has received considerable attention, not much work is done on the similar problem for reliability estimation. The main difficulty here is that linear combination of components of order statistics are good estimators of μ and σ but not of $R(t \mid \mu, \sigma)$. One could use the optimum estimators of μ and σ, say μ^* and σ^* based on components of order statistics, and substitute them in $R(t \mid \mu, \sigma)$ to obtain $R^*(t) = R(t \mid \mu, \sigma^*)$. However, not much is known about the effect of such a substitution but one could expect that $R^*(t)$ would have some advantages over the standard estimators, obtained earlier if the data contains outliers or spurious observations [Sinha (1972)].

EXERCISES

In the following exercises, unless otherwise stated, we will assume exponential distribution with mean life σ.

1.1 Suppose n items are put to test and the test is terminated (without replacement) as soon as the first $r(\leqslant n)$ failure times $x_{(1)} < x_{(2)} < \ldots < x_{(r)}$ are recorded.

    (i)   Obtain the maximum likelihood estimator of $1/\sigma$ and show that the estimator is biased,

    (ii) Find UMVUE of $1/\sigma$.

    (iii) Obtain the variances of these two estimators. Which of the two will you prefer?

1.2 In order to estimate the mean burning time of a particular brand of light bulb, 30 bulbs were left burning. The bulbs that failed were not replaced upon failure. The following burning times (in hours) were reported:

20, 27, 52, 61, 110, 122, 214, 232, 238, 371, 393, 426, 445, 472, 503, 526, 581, 627, 697, 798, 805, 909, 976, 1001, 1016, 1033, 1086, 1192, 1322, 1681.

Obtain the MLE of σ, the MLE and UMVUE of $R(t \mid \sigma)$ at $t = 900$ hours.

1.3 100 electronic tubes of a certain type were tested. The test terminated after the first 15 tubes blew. The failures occurred after the following hours:

    40, 60, 90, 120, 195, 260, 350, 420, 501, 620, 650, 730, 815, 910, 980.

Estimate σ and $R(t \mid \sigma)$ at $t = 500$ hours, assuming:

    (i)   the failed tubes are not replaced,

    (ii) the failed tubes are replaced by new tubes.

1.4 100 units were tested (without replacement) for 600 hours and the first 10 failures occurred after the following hours: 40, 50, 80, 90, 120, 190, 220, 290, 350, 460. Estimate σ by using:

    (i)   the failure times,

    (ii) the number of failures during the interval $(0, 600)$.

Which of the two estimators would you prefer to use? Why?

1.5 25 devices were tested, without replacement, for 100 hours. During this period 10 failures were reported after 5, 18, 24, 33, 42, 49, 53, 61, 70 and 90 hours. Using these failure times, obtain the MLE of σ and $R(t \mid \sigma)$, at $t = 50$ hours.

1.6 50 devices are under test without replacement. The test will be terminated at the instant when either the accumulated time on the test equals 5500 hours or when the 7th device fails and the total time when this happens < 5500 hours. Suppose the first four failures occurred after 50, 65, 85 and 100 hours.

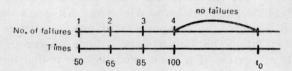

Compute the MLE of σ.

1.7 Consider a linear estimator $\sum\limits_{i=1}^{n} l_i x_i$ for σ. Show that $E(\Sigma l_i x_i - \sigma)^2 \geqslant \dfrac{\sigma^2}{n+1}$ with the equality attained for each $l_i = \dfrac{1}{n+1}$. The estimator $\dfrac{1}{n+1}\sum\limits_{i=1}^{n} x_i$ is known as the minimum mean squared estimator within the class of linear estimators $\sum\limits_{i=1}^{n} l_i x_i$.

1.8 Consider the first m order statistics $x_{(1)}, x_{(2)}, \ldots, x_{(m)}$ and estimators of the type

$\sum\limits_{i=1}^{m} w_i x_{(i)}$. Show that there is a one to one correspondence between $\sum\limits_{i=1}^{m} w_i x_{(i)}$ and

$\sum\limits_{i=1}^{m} c_i z_i$, where $z_i = (n - i + 1)(x_{(i)} - x_{(i-1)})$, $i = 1, 2, \ldots, m$, $x_{(0)} = 0$. Using

the fact that $(z_i, z_2, \ldots, z_m)$ are i.i.d. exponential r.v. with mean life σ, show

that $\dfrac{1}{m+1} \left\{ \sum\limits_{i=1}^{m} x_{(i)} + (n - m)x_{(m)} \right\}$ is minimum mean squared estimator with-

in the class of estimators of the type $\sum\limits_{i=1}^{m} w_i x_{(i)}$ (Kale and Sinha, 1972).

1.9 n_1 yellow and n_2 blue Christmas bulbs were put on test. The test was terminated as soon as the first $r (\leqslant n_1)$ blue bulbs failed. If the failure times are i.i.d. as $f(x, \sigma) = (1/\sigma) \exp (-x/\sigma)x$, $\sigma > 0$, obtain the m.l.e. of σ when the (blue) bulbs that failed:
   (i)   were not replaced,
   (ii) were replaced by new (blue) bulbs.

1.10 Suppose 50 motors of a particular kind are under test and the test is terminated after the first 15 motors failed. The first failure occurred 900 hours after the test started and the total life time between the first and the fifteenth failure is 15,000 hours. If the life distribution is a two-parameter exponential:

$$f(x \mid \mu, \sigma) = \frac{1}{\sigma} \exp \left\{ -\frac{x - \mu}{\sigma} \right\}, -\infty < \mu < x, \sigma > 0,$$

obtain UMVUE of σ and μ.

1.11 20 items under test failed at the following hours:
        10.99, 15.79, 24.14, 34.43, 43.72, 51.72, 56.12, 68.27, 77.20, 88.47, 91.07,
        117.58, 130 40, 133.12, 152.90, 159.00, 193.62, 208.71, 308.82, 316.07.
   Let the failure time distribution be given by:

$$f(x \mid \mu, \sigma) = \frac{1}{\sigma} \exp \left\{ -\left(\frac{x - \mu}{\sigma} \right) \right\}, x \geqslant \mu, \sigma > 0.$$

Compute the MLE and UMVUE of $R(t \mid \mu, \sigma)$ at $t = 100$ hours based on
   (a)   all the 20 observations,
   (b)   the first 15 observations.

CHAPTER 2

Gamma and Weibull Distributions

In Chapter 1, we considered the exponential distribution which is perhaps, the simplest and most often used distribution in life testing experiments and reliability theory. In this chapter we will consider the gamma and Weibull distributions as models for the failure time distribution of the items under test.

First, consider the two-parameter gamma distribution with p.d.f. given by

$$f(x \mid \sigma, p) = \frac{1}{\Gamma(p)} \frac{x^{p-1}}{\sigma^p} \exp\left(\frac{-x}{\sigma}\right), \quad x \geqslant 0, \ \sigma > 0, \ p > 0 \tag{1}$$

where $\Gamma(p)$ is the well-known gamma function which for integral values of p equals $(p - 1)!$. Note that for $p = 1$, the gamma distribution reduces to the one-parameter exponential distribution with parameter σ. As mentioned in Introduction, there are no simple closed expressions available for $F(t)$, $R(t)$ or $\mu(t)$ for the gamma distribution. However, $R(t)$ and $F(t)$ have been extensively studied and tabulated. We refer to Biometrika Tables for Statisticians, Vol. I and II by Pearson and Hartley (1954), Abramowitz and Stegun (1964) and further reference contained therein. The 'instantaneous failure rate', $\mu(t)$, can be shown to be an increasing function of t, indicating the 'aging effect', i.e., the failure or the hazard rate increases with the time (age t). In exponential distribution the hazard rate is constant $(= 1/\sigma)$, and, therefore, the gamma distribution immediately provides a generalization of the exponential distribution in this direction. The gamma distribution is one of the important distributions in statistical theory and arises in quite a few diverse contexts and the results derived in this chapter are useful in these contexts.

For the integral value of p, we know that the gamma distribution arises as a sum of p independent identically distributed exponential random variables. Therefore, if p items were on test and it was assumed that the failure time distribution is exponential with parameter σ, then the total time on test (or total of life times) would be a gamma variable with parameters p and σ. In failure censored case, with the experiment terminating at the $(p - r)$th failure, the total time on test would be distributed as gamma with parameters $(p - r)$ and σ. We can consider the following scheme called 'sliding standby' parts which leads to gamma distribution from the exponential distribution. Suppose that a life of an item depends on whether or not a given part is functioning. Suppose further that a substitute identical part immediately comes into operation as soon as the operating part fails. On the assumption that the failure time distribution of the part under consideration is exponential with parameter σ, it is easy to see that the failure time distribution of the item would be a sum of two independent exponential r.v.s., which is a gamma distribution with parameters σ and $p = 2$. In case there are p parts which are stored as 'sliding standby' and come into operation sequentially one at a time after the failure of the currently operating part, then it is clear that the failure time distribution of the item under consideration is the sum of p independent exponential r.v.s., which is a gamma with parameters σ and p. In this case, p is known and σ would be the only unknown parameter.

Estimation with Complete Sample

Suppose n items are subject to test and the test is terminated after all the items have failed. Let $(X_1, X_2, \ldots, X_n)$ be the random failure times and suppose that the failure times are distributed as gamma with parameters p and σ. The likelihood of the sample then is given by

$$L(x_1, \ldots, x_n \mid p, \sigma) = \frac{1}{[\Gamma(p)]^n \sigma^{np}} \exp \left\{ -\frac{\sum_{i=1}^{n} x_i}{\sigma} \right\} \prod_{i=1}^{n} x_i^{p-1}$$

We will first consider the case when p is known and the only unknown parameter is σ. By simple differentiation it can be shown that the MLE of σ is given by $\hat{\sigma} = \left(\sum_{i=1}^{n} x_i \right) / np$. We can easily verify that $E(\hat{\sigma}) = \sigma$ and var $(\hat{\sigma}) = \sigma^2/np$. The distribution of $\hat{\sigma}$ is gamma with parameter np and σ/np. This follows from the fact that $(X_1, X_2, \ldots, X_n)$ are i.i.d. gamma r.v. with parameters p, σ and, therefore, $\sum_{i=1}^{n} x_i$ is gamma with parameters np and σ and $\hat{\sigma} = \sum x_i/np$ is gamma with parameters np and σ/np. Thus, the p.d.f. of $\hat{\sigma}$ is given by

$$g(\hat{\sigma} \mid p, \sigma) = \frac{1}{\Gamma(np)} \frac{1}{(\sigma/np)^{np}} \exp \left\{ -\frac{\hat{\sigma}}{\sigma/np} \right\} \cdot (\hat{\sigma})^{np-1} \qquad (2)$$

From (1) and (2) it immediately follows

$$\frac{L(x_1, \ldots, x_n \mid p, \sigma)}{g(\hat{\sigma} \mid p, \sigma)} = \frac{\Gamma(np)}{[\Gamma(p)]^n} \frac{1}{(np)^{np}} \frac{\prod_{i=1}^{n} x_i^{p-1}}{\hat{\sigma}^{np-1}}$$

which is independent of the unknown parameter σ. Thus, $\hat{\sigma}$ (or equivalently the total life time $\sum_{i=1}^{n} x_i$) is sufficient for σ. The completeness of $\hat{\sigma}$ can be proved by using the uniqueness property of the Laplace transform and we refer to Lehmann and Scheffe (1955) for details. Thus it follows that $\hat{\sigma}$ is not only MLE but is also UMVUE of σ. In the gamma model we note that the mean life is $p\sigma$ and if we are interested in estimating $p\sigma$, then the MLE and UMVUE of $p\sigma$ are identical and are given by the sample mean $\frac{1}{n}\sum\limits_{i=1}^{n} x_i$.

Next, we consider the case when p and σ are both unknown. The likelihood of the sample is given by (2) and

$$\log L(x\,|\,p,\,\sigma) = -n \log \Gamma(p) - np \log \sigma + (p-1) \sum_{i=1}^{n} \log x_i - \frac{1}{\sigma} \sum_{i=1}^{n} x_i$$

This leads to the likelihood equations

$$\frac{\partial \log L}{\partial \sigma} = -\frac{np}{\sigma} + \frac{n\bar{x}}{\sigma^2} = 0 \tag{3}$$

$$\frac{\partial \log L}{\partial p} = -n\frac{d}{dp}[\log \Gamma(p)] - n \log \sigma + \sum_{i=1}^{n} \log x_i = 0 \tag{4}$$

The function $\frac{d}{dp}[\log \Gamma(p)]$ has no closed expression available and the above system of equations may be solved by an iterative procedure such as the Newton-Raphson process or the method of scoring for parameters. For details we refer to Kale (1962). Another method would be to use (3) to obtain $\hat{\sigma} = \bar{x}/p$ and substituting in (4) derive the equation

$$\frac{d}{dp} \log \Gamma(p) - \log p = \frac{1}{n} \sum_{i=1}^{n} \log x_i - \log \bar{x} \tag{5}$$

The MLE are $\hat{\sigma} = \bar{x}/\hat{p}$ where $\hat{p}$ is the solution of (5). The solution to (5) can be obtained by inverse interpolation methods. The function $\frac{d}{dp}\log \Gamma(p)$ is known as digamma function and is quite extensively tabulated in Abramowitz and Stegun (1964) and Pairman (1919). For large values of p, we can use the approximation

$$\frac{d}{dp}[\log \Gamma(p)] \simeq \log p - \frac{1}{2p} \quad \text{so that} \quad \hat{p} = \frac{1}{2\left[\log \bar{x} - \frac{1}{n}\sum\limits_{i=1}^{n} \log x_i\right]}$$

and $\hat{\sigma} = \bar{x}/\hat{p}$. We observe that the MLE of the mean life is again $\bar{x}$.

The exact joint distribution of $\hat{p}$ and $\hat{\sigma}$ when both parameters are unknown is extremely difficult to obtain. However, by using the general theory of maximum likelihood estimation, one can show that asymptotically $(\hat{p},\ \hat{\sigma})$ is distributed as bivariate normal with means p and σ respectively and the variance co-variance matrix is given by

$$\frac{1}{n}\begin{bmatrix} \dfrac{p}{\varDelta(p)} & \dfrac{-\sigma}{\varDelta(p)} \\[2ex] \dfrac{-\sigma}{\varDelta(p)} & \dfrac{-\psi'(p)\sigma^2}{\varDelta(p)} \end{bmatrix}$$

where $\varDelta(p) = p\psi'(p) - 1$ and $\psi'(p) = \frac{d^2}{dp^2}\{\log \Gamma(p)\}$ is the tri-gamma function which is tabulated. On the other hand, the MLE of the mean life $p\sigma$ is $\bar{x}$ and its distribution is gamma with parameters np and σ/n. For large samples the asymptotic distribution of $\bar{x}$ is normal with mean σ and variance $p\sigma^2/n$.

To prove the joint sufficiency of $(\hat{p}, \hat{\sigma})$ for (p, σ) is not easy. Indeed, even in not too complex models as gamma and Weibull, the general problem of obtaining complete sufficient statistics is difficult. Further, as the models become more complex such as three-parameter gamma distribution with p.d.f.

$$f(x \mid \mu, \sigma, p) = \frac{1}{\Gamma(p)} \frac{1}{\sigma^p} (x - \mu)^{p-1} \exp\left\{-\left(\frac{x - \mu}{\sigma}\right)\right\}, \quad x \geqslant \mu \quad (6)$$

to obtain complete sufficient statistics for (μ, σ, p) is very difficult. On the other hand, we know that the MLE's are functions of sufficient statistics and in most cases lead to estimators with desirable properties, at least in large samples. Therefore, we will not emphasize UMVUE approach in general and in this chapter in particular. We note that the MLE of (μ, σ, p) in the three-parameter gamma family are given by $\hat{\mu} = x_{(1)} = \mathrm{Min}(x_1, \ldots, x_n)$ and the $\hat{\sigma}$ and $\hat{p}$ are the solutions of the likelihood equations (3) and (4) where each x_i is replaced by $x_i - x_{(1)}$.

We note that in two-parameter gamma distribution, to obtain MLE we require the tables of digamma functions. The method of moments avoids this and leads to estimators $\hat{\hat{\sigma}}$ and $\hat{\hat{p}}$ which are easy to obtain. The method of moments consists of equating a suitable set of sample moments equal to the population moments. In the gamma distribution the mean and variance are given by $p\sigma$ and $p\sigma^2$ respectively. Therefore, the moment equations are

$$\bar{x} = p\sigma \quad (7)$$

$$\frac{1}{n-1} \sum_{i=1}^{n} (x_i - \bar{x})^2 = p\sigma^2 \quad (8)$$

which leads to

$$\hat{\hat{p}} = \frac{\bar{x}}{\hat{\hat{\sigma}}} \quad \text{and} \quad \hat{\hat{\sigma}} = \sum_{i=1}^{n} (x_i - \bar{x})^2/(n-1)\bar{x} \quad (9)$$

These estimators are, however, not as efficient as the MLE's $\hat{p}$ and $\hat{\sigma}$ in the sense that in large samples $\hat{\hat{p}}$ and $\hat{\hat{\sigma}}$ have larger variances than $\hat{p}$ and $\hat{\sigma}$ respectively. Note that if p is known the method of moments leads to $\hat{\hat{\sigma}} = \bar{x}/p$ (same as MLE). In any case, the MLE and the method of moments both lead to $\bar{x}$ for the mean life of the distribution. We refer to Kendall and Stuart (1972), Vol. II, pp. 66–67 for further details.

EXAMPLE 2.1   A random sample of size 25 was generated from the p.d.f. (1) (with $p = 1$, $\sigma = 10$). Compute the moment as well as the maximum likelihood estimators of p, σ.

SAMPLE

8.54415	0.35367	1.40631	23.30200	4.58690
8.83767	6.16266	7.83901	4.61845	23.20425
8.06476	5.86675	0.87095	0.99845	1.04360
6.85619	7.37164	9.32164	2.76115	6.53886
4.33080	1.48420	8.91998	8.91998	43.06764

$$n = 25, \ \bar{x} = 8.21087, \ \frac{\Sigma(x_i - \bar{x})^2}{n-1} = 86.11735$$

From (9) we have the moment estimators

$$\hat{\hat{\sigma}} = 10.4882, \ \hat{\hat{p}} = 0.7829, \ \sum_1^n \log x_i = 39.658. \quad \text{From (5)}$$

$$\psi(p) \equiv \frac{d}{dp} \log \Gamma(p) - \log p = 0.519$$

Chapman (1956) has tabulated $\psi(p)$ for a wide range of values of p. Using Chapman's table, by linear interpolation, we obtain

$$\hat{p} = 1.1024, \quad \hat{\sigma} = \bar{x}/\hat{p} = 7.448$$

	p	σ
Maximum likelihood estimators	1.1024	7.4482
Moment estimators	0.7829	10.4882
True values	1.0000	10.0000

Truncated and Censored Samples
Cohen (1950) estimated the mean and standard deviation of a singly truncated gamma distribution by the method of moments. Des Raj (1953) extended Cohen's results for singly and doubly truncated samples with known truncation points where the number of unmeasured observations is unknown for each tail, known separately for each tail and known jointly for the two tails. Den Broeder (1955) derived the MLE for α for the family

$$f(x \mid \alpha, p) = \frac{1}{\Gamma(p)} \alpha^p x^{p-1} \exp(-\alpha x), \quad x, p, \alpha > 0 \qquad (10)$$

where p is known and the sample is truncated at a known point $T > 0$. Chapman (1956) derived a technique similar to least square estimation in a truncated three-parameter gamma distribution (6) with known origin. Gupta and Groll (1961) discussed the use of gamma distribution in acceptance sampling with time-censored samples and presented a number of tables and charts for estimating the sample size necessary to guarantee a specified mean life, producer's risk, confidence interval, etc.

Wilk, Gnanadesikan and Huyett (1962) derived the MLE of α and p based on the order statistics $(x_1 < x_2 < x_3 < \ldots < x_r \mid r \leqslant n)$ from (10) where n is the size of the sample. The likelihood

$$L \propto \frac{\alpha^{rp}}{\{\Gamma(p)\}^r} \left[\prod_1^r x_i^{p-1} \right] \left[\exp\left(-\alpha \sum_1^r x_i \right) \right] \left[\frac{\alpha^p}{\Gamma(p)} \int_{x_r}^\infty x^{p-1} \exp(-\alpha x) \, dx \right]^{n-r}$$

$$\propto \frac{\alpha^{np}}{\{\Gamma(p)\}^{np}} \left[\prod_1^r x_i^{p-1} \right] \left[\exp\left(-\alpha \sum_1^r x_i\right) \right] \left[\int_{x_r}^{\infty} x^{p-1} \exp(-\alpha x)\, dx \right]^{n-r}$$

$$\text{Let } k = \alpha x_r, \quad A = \frac{\sum_1^r x_i}{r x_r}, \quad G = \frac{\left(\prod_1^r x_i\right)^{1/r}}{x_r}, \quad 0 \leqslant G \leqslant A \leqslant 1$$

$$L \propto \frac{k^{np}}{\{\Gamma(p)\}^n} \frac{G^{r(p-1)}}{x_r^r} \exp(-krA) J^{n-r}$$

where $J \equiv J(k, p) = \displaystyle\int_1^{\infty} u^{p-1} \exp(-ku)\, du$

$$\log L = \text{constant} + np \log k - n \log \Gamma(p) + r(p-1) \log G$$
$$- rAk + (n-r) \log J$$

$$\frac{\partial}{\partial p} \log L = n \log k - \frac{n \Gamma'(p)}{\Gamma(p)} + r \log G + \frac{(n-r) J'}{J}$$

where $\quad \Gamma'(p) = \dfrac{\partial}{\partial p} \Gamma(p), \ J' = \dfrac{\partial J}{\partial p} = \displaystyle\int_1^{\infty} u^{p-1} (\log u) \exp(-ku)\, du$

$$\frac{\partial}{\partial p} \log L = 0 \text{ leads to}$$

$$\log G = \frac{n \Gamma'(p)}{r \Gamma(p)} - \frac{n \log k}{r} - \left(\frac{n}{r} - 1\right) \frac{J'}{J}$$

Similarly, $\dfrac{\partial}{\partial k} \log L = 0$ yields

$$A = \frac{p}{k} - \frac{1}{k}\left(\frac{n}{r} - 1\right) \frac{\exp(-k)}{J}$$

Given G and A, the maximum likelihood estimators $\hat{k}, \hat{p}$ are solutions of the above equations. Wilk, Gnanadesikan and Huyett employed an elaborate iterative scheme of inverse interpolation to obtain k and p as functions of A and G.

Harter and Moore (1967) derived expressions for the elements of the information matrix for the maximum likelihood estimators of the parameters of a three-parameter gamma distribution (6).

Weibull Distribution

In this section we will consider the Weibull distribution as a model for the failure time distribution. The p.d.f. of the Weibull distribution is given by

$$f(x \mid \sigma, p) = \frac{p}{\sigma} x^{p-1} \exp\left\{-\frac{x^p}{\sigma}\right\}, \ x > 0, \ p > 0, \ \sigma > 0. \tag{11}$$

Here p is referred to as a shape parameter and σ as a scale parameter. Although we will follow this practice, we note that $\sigma' = \sigma^{1/p}$ should be regarded as the scale parameter.

As mentioned in Introduction, the Weibull distribution arises from the exponential distribution in the following way. Suppose that instead of assuming that the failure time is distributed exponentially, we assume some power (say pth) of the failure time is distributed exponentially. Thus, if X is the r.v. representing the failure time, then we assume that $Y = X^p$ has the p.d.f. given by

$$g(y, \sigma) = \frac{1}{\sigma} \exp\left(-\frac{y}{\sigma}\right), y \geqslant 0, \sigma > 0$$

Then by using the standard transformation technique, we can immediately show that the p.d.f. of X is given by (11) above.

For the Weibull distribution, the d.f. is given by $F(t) = 1 - \exp\{-t^p/\sigma\}$, the reliability function $R(t) = \exp\{-t^p/\sigma\}$ and the instantaneous failure rate or the hazard rate $\mu(t) = \frac{p}{\sigma} t^{p-1}$. For $p > 1$, $\mu(t)$ is increasing, for $p < 1$, $\mu(t)$ is decreasing and for $p = 1$, $\mu(t) = 1/\sigma$ is constant and leads to the exponential distribution. This observation leads to another way of deriving Weibull distribution; namely, assuming that the hazard rate $\mu(t)$ is proportional to a power of t. Thus, if we assume $\mu(t) = ct^k$ then, since $f(x) = \mu(x) \exp\left\{-\int_0^x \mu(w)\,dw\right\}$, we immediately obtain the corresponding p.d.f. as

$$f(x, c, k) = cx^k \exp\left\{-c\,\frac{x^{k+1}}{k+1}\right\}$$

which agrees with (11) if we take $c = p/\sigma$ and $k = (p - 1)$.

Weibull distribution has been extensively used in life testing and reliability problems. The distribution has been named after the Swedish scientist, Weibull, who proposed it for the first time in 1939 in connec-

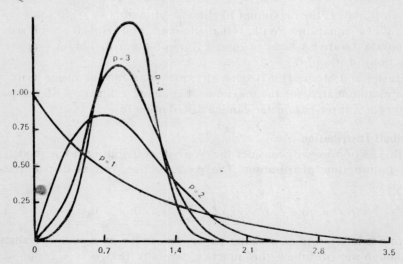

Fig. 1   Weibull density function with different
values of shape parameter, p.

tion with his studies on strength of material. Weibull (1951) showed that the distribution is also useful in describing the 'wear-out' or fatigue failures. Kao (1959) used it as a model for vacuum tube failures while Leiblin and Zelen (1956) used it as a model for ball bearing failures. Mann (1968) gives a variety of situations in which the distribution is used for other types of failure data.

In Figure 1, we have plotted Weibull density for different values of shape parameter p and $\sigma = 1$. As p becomes large the density gets more peaked and symmetric around its mean given by $\Gamma(1 + 1/p) \simeq 1$ for large values of p. As the graphs indicate even at $p = 4$ the symmetry around $\Gamma(1.25) = 0.9064$ is quite evident.

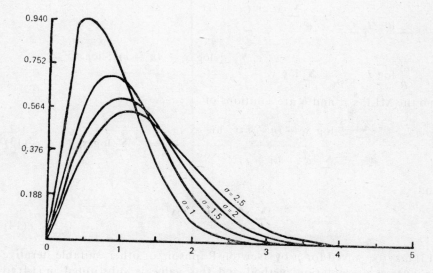

Fig. 2 Weibull density function with different
values of parameter, σ.

Figure 2 illustrates the graphs of the Weibull densities for $p = 2$ and different values of σ. We note that as σ gets larger the density gets less peaked and more asymmetric and having a rather long tail to the right.

Estimation of Parameters

We will first consider the case when both the parameters p and σ are unknown. Further, we will discuss the failure censored sample case where the experiment is terminated after r failures. As has been indicated earlier, the results for complete sample can be obtained by putting $r = n$ (Cohen, 1965). Let us, therefore, assume that we have a failure censored sample with the failure times of r items that failed. Let $x_{(1)} \leqslant x_{(2)} \ldots \leqslant x_{(r)}$ denote the failure times arranged in the increasing order of magnitude; then the likelihood may be written as follows, with $x_{(i)}$ replaced by x_i for typographical convenience.

$$L \propto \left(\frac{p}{\sigma}\right)^r \left(\prod_1^r x_i^{p-1}\right) \exp\left(-\frac{\sum_1^r x_i^p}{\sigma}\right) \left[\int_{x_r}^{\infty} \frac{p}{\sigma} x^{p-1} \exp\left(-\frac{x}{\sigma}\right) dx\right]^{n-r}$$

$$\propto \left(\frac{p}{\sigma}\right)^r \left(\prod_1^r x_i^{p-1}\right) \exp\left[-\frac{1}{\sigma}\left\{\sum_1^r x_i^p + (n-r)x_r^p\right\}\right] \tag{12}$$

$$\log L = \text{constant} + r\left(\log p - \log \sigma\right) + (p-1)\sum_1^r \log x_i$$

$$- \frac{\sum_1^r x_i^p + (n-r)x_r^p}{\sigma}$$

$$\frac{\partial}{\partial \sigma} \log L = -\frac{r}{\sigma} + \frac{\sum_1^r x_i^p + (n-r)x_r^p}{\sigma^2}$$

$$\frac{\partial}{\partial p} \log L = \frac{r}{p} + \sum_1^r \log x_i - \frac{\sum_1^r x_i^p \log x_i + (n-r)x_r^p \log x_r}{\sigma}$$

and the MLE's $\hat{p}$ and $\hat{\sigma}$ are solutions of

$$\frac{\sum_1^r x_i^{\hat{p}} \log x_i + (n-r)x_r^{\hat{p}} \log x_r}{\sum_1^r x_r^{\hat{p}} + (n-r)x_r^{\hat{p}}} - \frac{1}{\hat{p}} = \frac{1}{r}\sum_1^r \log x_i \tag{13}$$

and

$$\hat{\sigma} = \frac{\sum_1^r x_i^{\hat{p}} + (n-r)x_r^{\hat{p}}}{r} \tag{14}$$

(13) may be solved for $\hat{p}$ by Newton-Raphson or other suitable iterative or inverse interpolation method and this value is substituted in (14) to obtain $\hat{\sigma}$.

The exact distribution of $\hat{p}$ and $\hat{\sigma}$ is not known explicitly. However, by using the general theory of MLE, Cohen (1965) showed that $(\hat{p}, \hat{\sigma})$ is bivariate normal with mean (p, σ) and variance co-variance matrix given by

$$\begin{bmatrix} E\left(-\dfrac{\partial^2 \log L}{\partial p^2}\right) & E\left(-\dfrac{\partial^2 \log L}{\partial p\,\partial\sigma}\right) \\[2mm] E\left(-\dfrac{\partial^2 \log L}{\partial\sigma\,\partial p}\right) & E\left(-\dfrac{\partial^2 \log L}{\partial\sigma^2}\right) \end{bmatrix}^{-1}$$

The exact expressions for various expectations above are difficult to obtain. However, in practice we would need the estimate for the variance co-variance matrix which Cohen (1965) recommended using the approximations

$$E\left(-\frac{\partial^2 \log L}{\partial p\,\partial \sigma}\right) = -\frac{\partial^2 \log L}{\partial p\,\partial \sigma}\bigg|_{\substack{p=\hat{p} \\ \sigma=\hat{\sigma}}}, \quad \text{etc.}$$

These are given below.

$$-\frac{\partial^2 \log L}{\partial p^2}\bigg|_{\hat{p},\,\hat{\sigma}} = \frac{r}{\hat{p}^2} + \frac{\sum_1^r x_i^{\hat{p}}(\log x_i)^2 + (n-r)x_r^{\hat{p}}(\log x_r)^2}{\hat{\sigma}}$$

$$-\frac{\partial^2 \log L}{\partial p\,\partial \sigma}\bigg|_{\hat{p},\,\hat{\sigma}} = -\frac{\sum_1^r x_i^{\hat{p}} \log x_i + (n-r)x_r^{\hat{p}} \log x_r}{\hat{\sigma}^2}.$$

$$\frac{\partial^2 \log L}{\partial \sigma^2}\bigg|_{\hat{p},\,\hat{\sigma}} = -\frac{r}{\hat{\sigma}^2} + \frac{2}{\hat{\sigma}^3}\left\{\sum_1^r x_i^{\hat{p}} + (n-r)x_r^{\hat{p}}\right\}$$

We next consider the moment estimators. One can easily show that the mean and the variance of the Weibull distribution are given by

$$\mu_1' = \Gamma\left(\frac{p+1}{p}\right)\sigma^{1/p}$$

and

$$\mu_2 = \sigma^{2/p}\left[\Gamma\left(\frac{p+2}{p}\right) - \Gamma^2\left(\frac{p+1}{p}\right)\right]$$

(15)

The coefficient of variation, V_p, is independent of σ and is given by

$$\sqrt{\frac{\mu_2}{\mu_1'^2}} = \sqrt{\frac{\Gamma((p+2)/p) - \Gamma^2((p+1)/p)}{\Gamma^2((p+1)/p)}}$$

Table 2.1 Shape Parameter p Corresponding to the Coefficient of Variation c

c	p	c	p	c	p
429.8314	0.10	0.7238	1.40	0.3994	2.70
47.0366	0.15	0.7006	1.45	0.3929	2.75
15.8430	0.20	0.6790	1.50	0.3866	2.80
8.3066	0.25	0.6588	1.55	0.3805	2.85
5.4077	0.30	0.6399	1.60	0.3747	2.90
3.9721	0.35	0.6222	1.65	0.3690	3.00
3.1409	0.40	0.6055	1.70	0.3634	3.05
2.6064	0.45	0.5897	1.75	0.3581	3.10
2.2361	0.50	0.5749	1.80	0.3529	3.15
1.9650	0.55	0.5608	1.85	0.3479	3.20
1.7581	0.60	0.5474	1.90	0.3430	3.25
1.5948	0.65	0.5348	1.95	0.3383	3.30
1.4624	0.70	0.5227	2.00	0.3336	3.35
1.3529	0.75	0.5112	2.05	0.3292	3.40
1.2605	0.80	0.5003	2.10	0.3248	3.45
1.1815	0.85	0.4898	2.15	0.3206	3.50
1.1130	0.90	0.4798	2.20	0.3165	3.55
1.0530	0.95	0.4703	2.25	0.3124	3.60
1.0000	1.00	0.4611	2.30	0.3085	3.65
0.9527	1.05	0.4523	2.35	0.3047	3.70
0.9102	1.10	0.4438	2.40	0.3010	3.75
0.8718	1.15	0.4341	2.45	0.2974	3.80
0.8369	1.20	0.4279	2.50	0.2938	3.85
0.8050	1.25	0.4204	2.55	0.2904	3.90
0.7757	1.30	0.4131	2.60	0.2870	3.95
0.7487	1.35	0.4062	2.65	0.2838	4.00

Table 2.1 gives values for V_p which we can use to obtain the estimator p^* by equating $V_p = s/\bar{x}$, where $\bar{x}$ is sample mean and $s^2 = \dfrac{1}{n-1} \Sigma(x_i - \bar{x})^2$. From (15) we can immediately obtain the moment estimator of σ, say σ^*, given by

$$\sigma^* = \left[\frac{\bar{x}}{\Gamma\left(\dfrac{p^*+1}{p^*}\right)}\right]^{p^*} \tag{16}$$

We also note that p^* can be used as a starting value to obtain $\hat{p}$ by some iterative procedures or by inverse interpolation methods. A numerical example given below would illustrate these points. We also remark here that the results for time censored samples can be obtained by replacing x_r by t_0, the time the experiment was terminated (Cohen, 1965).

EXAMPLE 2.2   A random sample of 25 observations was generated from the Weibull distribution with $p = 2$, $\sigma = 4$.
  (i) Obtain the moment estimators (p^*, σ^*).
  (ii) Obtain the ML estimators $(\hat{p}, \hat{\sigma})$ and estimate the variance-covariance matrix.

SAMPLE

1.8487	0.3761	0.7500	3.0530	1.3545
1.8802	1.5700	1.7708	1.3592	3.0466
1.7961	1.5319	0.5903	0.6288	0.6461
1.6560	1.7172	1.9310	1.0509	1.6173
1.3162	0.7705	1.8889	1.8889	4.1505

  (i)   We have $\bar{x} = 1.607$, $s = 0.8538$, $s/\bar{x} = 0.5313$. From Table 2.1, $p^* = 1.964$ and from (16), $\sigma^* = 3.216$.

  (ii)   We have $r = n = 25$. We rewrite $\dfrac{\partial}{\partial p} \log L = 0$, as $g(p) = 0$, where

$$g(p) = \frac{\sum\limits_{1}^{25} x_i^p \log x_i}{\sum\limits_{1}^{25} x_j^p} - \frac{1}{p} - \frac{1}{25}\sum\limits_{1}^{25} \log x_i$$

We note that $g(2.05) > 0$ and so is $g(2.04)$; but $g(2.04) < g(2.05)$. So we calculate $g(p)$ for intermediary values of p.

p	$\sum\limits_{1}^{25} x_i^p$	$\sum\limits_{1}^{25} x_i^p \log x_i$	$g(p)$
1.9	75.7569	60.5168	−0.0620
2.0	82.1092	67.2588	−0.0157
2.025	83.8133	69.0694	−0.0045
2.035	84.5076	69.8085	−0.0001
2.03515	84.5181	69.8196	−0.0001
2.03525	84.5250	69.8271	0.0000
2.04	80.8576	70.1813	0.0021
2.05	85.5563	70.9333	0.0064

So $\hat{p} \approx 2.03525$ and

$$\hat{\sigma} \approx \frac{\sum\limits_{1}^{25} x_i^{\hat{p}}}{25} = 3.381$$

$$-\frac{\partial^2 \log L}{\partial p^2}\bigg|_{\hat{p},\hat{\sigma}} = \frac{n}{\hat{p}^2} + \frac{\sum\limits_{1}^{n} x_i^{\hat{p}} \log x_i}{\hat{\sigma}} = 28.02844$$

$$-\frac{\partial^2 \log L}{\partial p\,\partial\sigma}\bigg|_{\hat{p},\hat{\sigma}} = -\frac{1}{\hat{\sigma}^2}\sum\limits_{1}^{n} x_i^{\hat{p}} \log x_i = -6.108481$$

$$-\frac{\partial^2 \log L}{\partial\sigma^2}\bigg|_{\hat{p},\hat{\sigma}} = -\frac{n}{\hat{\sigma}^2} + \frac{2}{\hat{\sigma}^3}\sum\limits_{1}^{n} x_i^{\hat{p}} = 2.187003$$

$$\begin{bmatrix} 28.02844 & -6.108481 \\ -6.108481 & 2.187003 \end{bmatrix}^{-1} = \begin{bmatrix} 0.09117 & 0.2546 \\ 0.2546 & 1.1684 \end{bmatrix}$$

Thus

$$\mathrm{Var}\,(\hat{p}) = 0.09117$$

$$\mathrm{Var}\,(\hat{\sigma}) = 1.1684$$

and the coefficient of correlation between the two estimators

$$\rho_{\hat{p}\hat{\sigma}} = \frac{\mathrm{Cov}\,(\hat{p},\,\hat{\sigma})}{\sqrt{\mathrm{Var}\,(\hat{p})\,\mathrm{Var}\,(\hat{\sigma})}} = 0.78010$$

	True Value	Moment Estimator	M.L.E.
p	2	1.964	2.035
σ	4	3.216	3.381

Now consider the case where the parameter p is known. Following Harter and Moore (1965), we now consider the new r.v. $Y = X^p$ where p is known. The p.d.f. of Y is now exponential density and is given by

$$g(y, \sigma) = \frac{1}{\sigma} \exp\left\{-\frac{y}{\sigma}\right\}, \qquad y \geqslant 0, \sigma > 0$$

Assuming that we have a failure-censored sample with $y_1 < y_2 \ldots < y_r$ failure times, the MLE of σ is given by

$$\hat{\sigma} = \left\{\sum\limits_{i=1}^{r} y_i + (n-r)y_r\right\}\bigg/r$$

Again, the results for the complete sample case are obtained by putting $r = n$. As noted earlier, we know that σ is not really a scale parameter for the Weibull density given by (11), since the PDF can not be expressed as $(1/\sigma)f(x/\sigma)$. The scale parameter is $\sigma' = \sigma^{1/p}$.

Hence, we reparametrize (11) to obtain

$$f(x, b, c) = \left(\frac{c}{b}\right)\left(\frac{x}{b}\right)^{c-1} \exp\left\{-\left(\frac{x}{b}\right)^c\right\} \qquad x \geqslant 0, b > 0, c > 0 \qquad (17)$$

where $c = p$, $b^c = \sigma$ and b now is a scale parameter.

The MLE of b now is given by

$$\hat{b} = \left[\frac{\sum\limits_{i=1}^{r} x_i^c + (n-r)x_r^c}{r} \right]^{1/c} \tag{18}$$

and $\hat{b}^c = \hat{b}^p = \hat{\sigma}$. Since $r\hat{\sigma}/\sigma$ is a gamma variable with parameter r, the p.d.f. of $(\hat{b})^c$ is given by

$$f_1(\hat{b}^c) = \frac{1}{\Gamma(r)} \left\{ r\left(\frac{\hat{b}}{b}\right)^c \right\}^{r-1} \left(\frac{r}{b^c}\right) \exp\left\{ -r\left(\frac{\hat{b}}{b}\right)^c \right\}$$

$$= \frac{1}{\Gamma(r)} \left(\frac{r}{b^c}\right)^r (\hat{b}^c)^{r-1} \exp\left\{ -r\left(\frac{\hat{b}}{b}\right)^c \right\}, \qquad 0 < b^c < \infty$$

and the p.d.f. of $\hat{b}$ may be written as

$$f_2(\hat{b}) = \frac{c}{\Gamma(r)} \left(\frac{r}{b^c}\right)^r (\hat{b})^{cr-1} \exp\left\{ -r\left(\frac{\hat{b}}{b}\right)^c \right\}, \qquad 0 < b < \infty$$

$$E(\hat{b}) = \frac{c}{\Gamma(r)} \left(\frac{r}{b^c}\right)^r \int_0^\infty (\hat{b})^{cr} \exp\left\{ -r\left(\frac{\hat{b}}{b}\right)^c \right\} d_b$$

$$= \frac{1}{\Gamma(r)} \left(\frac{r}{b^c}\right)^r \int_0^\infty y^{r+(1/c)-1} \exp\left(-\frac{ry}{b^c}\right) dy$$

$$= \frac{b\,\Gamma\left(r + \dfrac{1}{c}\right)}{r^{1/c}\Gamma(r)} \tag{19}$$

Similarly,

$$E(\hat{b}^2) = \frac{b^2\Gamma\left(r + \dfrac{2}{c}\right)}{r^{2/c}\Gamma(r)}$$

and

$$\text{Var}\,(\hat{b}) = \frac{b^2}{r^{2/c}\Gamma(r)} \left[\Gamma\left(r + \frac{2}{c}\right) - \frac{\Gamma^2\left(r + \dfrac{1}{c}\right)}{\Gamma(r)} \right] \tag{20}$$

From (19) and (20), it follows that

$$\tilde{b} = \frac{r^{1/c}\Gamma(r)}{\Gamma\left(r + \dfrac{1}{c}\right)} \,\hat{b} = k(r,\,c)\hat{b} \tag{21}$$

is an unbiased estimator of b and

$$\text{Var}\,(\tilde{b}) = \frac{r^{2/c}\Gamma^2(r)}{\Gamma^2\left(r + \dfrac{1}{c}\right)} \,\text{Var}\,(\hat{b}) = b^2 \left[\frac{\Gamma(r)\Gamma\left(r + \dfrac{2}{c}\right)}{\Gamma^2\left(r + \dfrac{1}{c}\right)} - 1 \right] \tag{22}$$

Harter and Moore (1965) have tabulated $k(r, c)$, the correcting factor for bias and Var $\frac{(\tilde{b})}{b^2}$ for a wide range of values of r and c.

EXAMPLE 2.3 Based on the first 10 order statistics out of a total of 25 observations from the Weibull p.d.f. (17) with $c = 2$, compute the following:

(i) the MLE of b and an estimate of its standard error;

(ii) an unbiased estimate of b and an estimate of its standard errror; and

(iii) the efficiency of the unbiased estimator of b on the first 10 observations with respect to its maximum likelihood estimator based on the total observations.

SAMPLE

0.3761,   0.5902,   0.6288,   0.6461,   0.7500,

0.7705,   1.0509,   1.3162,   1.3545,   1.3592

We have $n = 25$, $r = 10$, $c = 2$.

(i) From (18) and (20), we obtain

$$\hat{b} = \left[\frac{8.97760 + 27.71137}{10} \right]^{1/2}$$

$$= 1.9154$$

and

$$\text{the estimate of the Var } (\hat{b}) = \frac{(1.9154)^2}{10\Gamma(10)} \left[\Gamma(11) - \frac{\Gamma^2(10.5)}{\Gamma(10)} \right]$$

$$= (1.9154)^2 \left[1 - \frac{\Gamma^2(10.5)}{\Gamma(10)\Gamma(11)} \right]$$

$$= 3.6689(1 - 3.12364 \times 0.312364)$$

$$= 0.0891$$

Thus, the estimate of the standard error of $\hat{b} = 0.2985$

(ii) From (21) and (22),

$$\tilde{b} = \text{unbiased estimator of } b$$

$$= \frac{\sqrt{10\Gamma(10)} \times 1.9154}{\Gamma^2(10.5)}$$

$$= 1.9391$$

The estimate of the Var $(\tilde{b}) = \frac{10\Gamma^2(10)}{\Gamma(10.5)} \times 0.0891$

$$= 0.0913$$

Thus the estimate of the standard error of $\tilde{b} = 0.3022$.

(iii) Let $V_r(t)$ denote the variance of the estimator t based on the first r observations. Using Harter and Moore (1965) Table 1, efficiency of $\tilde{b}$ is given by

$$\frac{V_{25}(\hat{b})}{V_{10}(\tilde{b})}$$

$$= \frac{0.01004949}{0.02530447}$$

$$= 39.7\%$$

Mendenhall and Lehman (1960) studied the mean and variance of the estimator of σ for the p.d.f. (11) where the shape parameter p is known and the testing is terminated after a pre-assigned time period t. The MLE of σ is given by

$$\hat{\sigma} = \frac{\sum_1^r x_i^p + (n - r)t_0^p}{r} \tag{23}$$

where $(x_1 < x_2 < \ldots < x_r)$ is a time-censored sample, r is the number of failures before t_0 and $(n - r)$ is the number surviving to time t_0. Mean and variance of $\hat{\sigma}$ are derived as functions of negative order moments of a binomial variable r assumed positive

$$E(\hat{\sigma}) = \sigma + t_0^p\left[nE\left(\frac{1}{r}\right) - \frac{1}{P}\right]$$

$$V(\hat{\sigma}) = t_0^{2p}n^2 V\left(\frac{1}{r}\right) + \left(\sigma^2 - \frac{Qt_0^{2p}}{P^2}\right)E\left(\frac{1}{r}\right) \tag{24}$$

where

$$Q = \exp\left(-\frac{t_0^p}{\sigma}\right), \quad P = 1 - Q \quad \text{and}$$

$$E\left(\frac{1}{r}\right) \approx \frac{n - 2}{n(a - 1)}$$

$$V\left(\frac{1}{r}\right) \approx \frac{(n - 2)(n - a - 1)}{n^2(a - 1)^2(a - 2)}, \quad a = (n - 1)P \tag{25}$$

$$E\left(\frac{1}{r^k}\right), \quad k > 0, \quad \text{are tabulated}$$

EXAMPLE 2.4   25 items were placed on test and the test was terminated after a pre-assigned time $t_0 = 20$ hours from the starting time. 10 items failed before time t_0 at 3.8, 5.9, 6.3, 6.5, 7.5, 7.7, 10.5, 13.2, 13.5 and 14.0 hours. If the underlying life distribution is given by (11) with $p = 2$, compute
  (i) $\hat{\sigma}$, the MLE of σ, and
  (ii) estimated standard error of $\hat{\sigma}$.
We have $n = 25$, $r = 10$, $t_0 = 20$. From (23),

$$\hat{\sigma} = \frac{909.47 + 15 \times 400}{10} = 690.947, \quad \hat{Q} = 0.56046$$

Using the estimates $\hat{Q}$ and $\hat{\sigma}$, from (25)

$$a = 10.54896,\ E(1/r) \approx 0.09635,\ V(1/r) \approx \frac{0.39688}{n^2} = 0.00064$$

Substituting in (24), the estimated Var $(\hat{\sigma}) = 64777$ and hence an estimate of the standard error of $\hat{\sigma} = 254.51$.

The problems of obtaining UMVUE for p and σ or any suitable function of (p, σ) will not be considered here. Not much work has been done in this area mainly due to the problems related to obtaining a complete sufficient statistics for (p, σ). On the other hand, if p is known then $\left\{\sum_{i=1}^{r} x_i^p + (n-r)x_r^p\right\}$ is a complete sufficient statistic for σ and $\hat{\sigma} = \left\{\sum_{i=1}^{r} x_i^p + (n-r)x_r^p\right\}/r$ is UMVUE of σ. This easily follows from the results obtained for the exponential distribution in Chapter I, noting that $Y = X^p$ is, in fact, exponential with parameter σ. As noted earlier, $b = \sigma^{1/p}$ is the scale parameter and in the case where p is known $\tilde{b}$ given by (21) is UMVUE of b, since $\tilde{b}$ can be expressed as a function of the complete sufficient statistics $\left\{\sum_{i=1}^{r} x_i^p + (n-r)x_r^p\right\}$.

In view of the difficulties involved in estimation problems for the two-parameter Weibull distribution, we consider some other methods of estimation. For a more extensive treatment we refer to Mann, Schafer and Singpurwalla (1974). We will consider the Weibull distribution as given by (17), namely,

$$f(x, b, c) = \frac{c}{b}\left(\frac{x}{b}\right)^{c-1} \exp\left\{-\left(\frac{x}{b}\right)^c\right\},\quad x \geqslant 0, b > 0, c > 0$$

Menon (1963) estimates $d = 1/c$ and $\log b$ instead of c and b due to some seemingly 'nice' properties of these estimators. Let $y = (x/b)^c$. From (17) we have

$$g(y) = \exp(-y),\quad y > 0$$

By definition,

$$\Gamma(p) = \int_0^\infty w^{p-1} \exp(-w)\, dw$$

Since gamma function is continuous and possesses continuous derivatives of all orders [Cramér (1961)], we have

$$\psi(p) = \frac{\partial}{\partial p} \log \Gamma(p)$$

$$= \frac{1}{\Gamma(p)}\int_0^\infty w^{p-1}(\log w) \exp(-w)\, dw$$

Differentiating once again,

$$\Gamma(p)\{\psi'(p) + \psi^2(p)\} = \int^\infty w^{p-1}(\log w)^2 \exp(-w)\, dw$$

where

$$\psi(p) = \frac{1}{p-1} + \psi(p-1), \quad p > 1$$

$$\psi(1) = -0.5772 \quad \text{(Euler's constant)}$$

$$\psi'(p) = \frac{\pi^2}{6} - \sum_{k=1}^{p-1} \frac{1}{k^2} \quad \text{if } p \text{ is an integer, and}$$

$$= \sum_{k=1}^{\infty} \frac{1}{(p+k-1)^2} \text{ otherwise} \quad \text{(Pairman, 1919)}$$

Hence,

$$E(\log y) = \int_0^\infty (\log y) \exp(-y)\, dy = \psi(1)$$

$$E(\log y^2) = \int_0^\infty (\log y)^2 \exp(-y)\, dy = \psi'(1) + \psi^2(1)$$

$$\text{Var}(\log y) = \psi'(1) = \pi^2/6$$

Also $y = (x/b)^c$, $\log y = c(\log x - \log b)$ and $\text{Var}(\log y) = c^2 \text{Var}(\log x)$. Thus

$$c^2 \text{Var}(\log x) = \pi^2/6$$

i.e.,
$$\hat{c}^2 \{\text{estimated Var}(\log x)\} = \pi^2/6$$

or

$$\frac{1}{\hat{c}} = \sqrt{\frac{6}{\pi^2} \left[\frac{\sum_1^n (\log x_i)^2 - \dfrac{\left\{\sum_1^n (\log x_i)\right\}^2}{n}}{n-1} \right]^{1/2}} = \hat{d} \qquad (26)$$

Also $E(\log y) = c[E(\log x) - \log b]$ and letting $d = 1/c$ we have

$$\psi(1)d = E(\log x) - \log b$$

and
$$\log b = E(\log x) - d\psi(1)$$

i.e.,
$$\widehat{(\log b)} = \frac{\sum_1^n (\log x_i)}{n} + 0.5772\hat{d} \qquad (27)$$

where $\hat{d}$ has been obtained in (26). $\hat{d}$ and $\widehat{(\log b)}$ are asymptotically unbiased and

$$\text{Var}(\hat{d}) \approx 1.1 \frac{d^2}{n} + d^2 0\left(\frac{1}{n^2}\right)$$

$$\text{Var}(\widehat{\log b}) \approx 1.2 \frac{d^2}{n} + d^2 0\left(\frac{1}{n^2}\right) \quad \text{(Cramér, 1961)}$$

For the data in Example 2.2,

$$\sum_1^{25} (\log x_i) = 8.40418$$

$$\sum_{1}^{25} (\log x_i)^2 = 10.45648$$

From (26)

$$\hat{d}^2 = \frac{6(10.45648 - 2.825210)}{9.868394 \times 24}$$

$$\hat{d} = 0.439688, \quad \hat{c} = \frac{1}{\hat{d}} = 2.773$$

and from (27)

$$\widehat{(\log b)} = 0.336174 + 0.5772 \times 0.439688$$

$$= 0.589955$$

This leads to $\hat{c}\,\widehat{(\log b)} = 1.34097$ and $\hat{\sigma} = 3.8229$.

We summarize the various estimators for p and σ for the data in Example 2.2

	$p = 2$	$\sigma = 4$
True value		
Moment estimators	1.964	3.216
MLE	2.035	3.381
Menon's method	2.273	3.823

Next, we will consider another method of estimating c and b due to White (1969) which is based on a regression approach. Unlike Menon's method, the method proposed by White is applicable to censored samples as well as complete samples. Let $t_1 < t_2 < \ldots < t_r$, $(r \leqslant n)$ denote the ordered failure times of the first r items and let us assume that the failure times follow the Weibull distribution with shape parameter c and scale parameter b, as given by (17). The corresponding d.f. is given by

$$F(t_i) = 1 - \exp\left\{-\left(\frac{t_i}{b}\right)^c\right\}$$

and taking logarithms we obtain

$$\log t_i = \log b + \frac{1}{c} \log \left[-\log (1 - F(t_i))\right]$$

which can be written as

$$y_i = \alpha + \beta x_i \tag{28}$$

where $y_i = \log t_i$, $\alpha = \log b$, $\beta = 1/c$ and $x_i = \log \{-\log (1 - F(t_i))\}$. Now here y_i are known and α and β are unknown parameters. If x_i were known, one could apply standard regression approach to estimate (α, β) and thus (b, c). However, although x_i are not known their distribution is known. Indeed, $(x_1, x_2, \ldots, x_r)$ are order statistics from the distribution with the density

$$h(w) = \exp\{w - e^w\}, \quad -\infty < w < \infty \tag{29}$$

The proof is based on the fact that if a r.v. U has a d.f. $F(u)$ then $V = [1 - F(U)]$ is uniformly distributed over $(0, 1)$ and $W = \log[-\log V]$ is distributed with p.d.f. given in (29). Note that this result is true for any continuous r.v. U with d.f. $F(u)$ and p.d.f. $f(u)$. White, therefore, suggests estimating x_i by the expected value of the ith order statistics (in a sample of size n), from (29) which is defined as the reduced log-Weibull distribution. White (1969) suggests comparing the observed $(y_1, y_2, \ldots, y_r)$ with $E(y_1), E(y_2), \ldots, E(y_r)$ by using the relation

$$E(Y_i) = \alpha + \beta E(X_i), \quad i = 1, 2, \ldots, r \leqslant n \tag{30}$$

White (1969) has tabulated $E(X_i)$, $1 \leqslant i \leqslant n \leqslant 20$. [For $20 \leqslant n \leqslant 100$, one may refer to the General Motors Publication GMR-717 or contact Dr. White.] Using (30), White obtains the least squares estimators $\hat{\alpha}$ and $\hat{\beta}$ which leads to the estimators

$$\tilde{b} = \exp(\hat{\alpha}) \quad \text{and} \quad \tilde{c} = 1/\hat{\beta}$$

We consider now a numerical example to illustrate the above method due to White.

EXAMPLE 2.5   A random sample of 20 observations from the Weibull density given by (17) is generated (with $b = 4$, $c = 2$). The first fifteen ordered observations are:

0.75684, 1.17739, 1.26504, 1.49810, 1.54122,

2.10067, 2.63326, 2.70912, 2.71897, 3.06323,

3.14009, 3.31184, 3.43607, 3.54193, 3.59613

The corresponding $E(X_{n,i})$, $n = 20$, $i = 1, 2, \ldots, 15$ as tabulated by White (1959) are:

$-3.57295, -2.54708, -2.02004, -1.65853, -1.37856,$

$-1.14714, -0.94723, -0.76907, -0.60638, -0.45483,$

$-0.31116, -0.17271, -0.03713, \quad 0.09791, \quad 0.23503$

We wish to obtain White's estimates of b and c.
We have a failure-censored sample where $n = 20$, $r = 15$:

$X = E(X_{n,i})$	$Y = \log t_i$
-3.57295	-0.27860
-2.54708	0.16330
-2.02004	0.23510
-1.65835	0.40420
-1.37856	0.43257

$X = E(X_{n,i})$	$Y = \log t_i$
-1.14714	0.74226
-0.94723	0.96822
-0.76907	0.99662
-0.60638	1.00025
-0.45483	1.11947
-0.31116	1.14425
-0.17271	1.19750
-0.03713	1.23433
0.09791	1.26467
0.23503	1.27986

$\Sigma xy = -4.99697, \ \Sigma x = -15.28987, \ \Sigma y = 11.904, \ \Sigma x^2 = 31.55736,$

$$\hat{\beta} = \frac{\Sigma xy - \dfrac{\Sigma x \Sigma y}{15}}{\Sigma x^2 - \dfrac{(\Sigma x)^2}{15}} \text{ and } \hat{\alpha} = \bar{y} - \hat{\beta}\bar{x} = 1.24908$$

Hence $\hat{c} = 1/\hat{\beta} = 2.238$, $\hat{b} = \exp(\hat{\alpha}) = 3.487$. The MLE are $\hat{c} = 2.593$, $\hat{b} = 3.303$.

We summarize the results of the above example,

	b	c
True value	4	2
White's method	3.487	2.238
MLE	3.303	2.593

Reliability Estimation

Let $R(t) = 1 - F(t \mid p, \sigma)$ be the reliability function. Then for gamma density with parameters p and σ, the $R(t \mid p, \sigma)$ is given by

$$R(t \mid p, \sigma) = \int_t^\infty \frac{1}{\Gamma(p)} \frac{x^{p-1}}{\sigma^p} \exp\left(-\frac{x}{\sigma}\right) dx$$

The MLE of $R(t \mid p, \sigma)$ is given by $R(t \mid \hat{p}, \hat{\sigma})$, where $\hat{p}$ and $\hat{\sigma}$ are MLE of p and σ respectively. Thus, for the complete sample case we have

$$\hat{R}(t \mid p, \sigma) = R(t \mid \hat{p}, \hat{\sigma})$$

$$= \frac{1}{(\hat{\sigma})^{\hat{p}} \Gamma(\hat{p})} \int_t^\infty x^{\hat{p}-1} \exp\left(-\frac{x}{\hat{\sigma}}\right) dx$$

$$= \frac{1}{\Gamma(\hat{p})} \int_{t/\hat{\sigma}}^\infty y^{\hat{p}-1} \exp(-y) \, dy$$

For a given value of t, $\hat{p}$ and $\hat{\sigma}$, the above integral can be evaluated by using the tables of incomplete gamma function [K. Pearson, 1968] or by numerical integration. For the data in the Example 2.1 we have the MLE of p and c are given by $\hat{p} = 1.1024$ and $\hat{\sigma} = 7.4482$ and the following table gives the MLE $\hat{R}(t \mid p, \sigma)$ for different values of t.

t	$\hat{R}(t \mid p, \sigma)$	True value
10	0.2999	0.3679
15	0.1582	0.2231
20	0.0829	0.1333
25	0.0433	0.0821

In the case where p is known, the MLE of $R(t \mid p, \sigma)$ would be $\hat{R}(t \mid p, \hat{\sigma})$ where $\hat{\sigma} = \bar{x}/p$. Thus, for the case when p is known

$$\hat{R}(t \mid p, \sigma) = \frac{1}{\Gamma(p)} \int_{tp/\bar{x}}^{\infty} y^{p-1} \exp(-y) \, dy$$

We have seen earlier that when p is known $\hat{\sigma} = \bar{x}/p$ is UMVUE of σ and $\hat{\sigma}$ can be shown to be a complete sufficient statistic for σ. In the case, when p is known, we can obtain UMVUE of $R(t \mid p, \sigma)$ in the same way as it was done for the exponential case $p = 1$. The first step is to show that the joint p.d.f. of X_1 and $\hat{\sigma}$ is given by

$$f(x_1, \hat{\sigma} \mid p, \sigma) = \frac{np(n-1)^{(n-1)p-1}}{\Gamma(p)\Gamma\{(n-1)p\}} \frac{x_1^{p-1}}{\sigma^{np}}$$

$$\times \left(\frac{np\hat{\sigma}}{n-1} - \frac{x_1}{n-1}\right)^{(n-1)p-1} \exp\left(-\frac{np\hat{\sigma}}{\sigma}\right), \quad 0 < x_1 < np\hat{\sigma}$$

We have also seen that the p.d.f. of $\hat{\sigma}$ is given by

$$g(\hat{\sigma} \mid p, \sigma) = \frac{1}{\Gamma(np)} \left(\frac{np}{\sigma}\right)^{np} (\hat{\sigma})^{np-1} \exp\left\{-\frac{np\hat{\sigma}}{\sigma}\right\}$$

which immediately leads to the conditional p.d.f. of X_1 given $\hat{\sigma}$ to be

$$f(x_1 \mid \hat{\sigma}, p) = \frac{\Gamma(np)}{\Gamma(p)\Gamma\{(n-1)p\}} \left\{1 - \frac{x_1}{np\hat{\sigma}}\right\}^{(n-1)p-1} \frac{x_1^{p-1}}{(np\hat{\sigma})^p}, \quad 0 < x_1 < np\hat{\sigma}$$

$$= \frac{1}{B\{p, (n-1)p\}} \frac{x_1^{p-1}}{(n\bar{x})^p} \left\{1 - \frac{x_1}{n\bar{x}}\right\}^{(n-1)p-1}, \quad 0 < x_1 < n\bar{x}$$

which shows that $W = x_1/n\bar{x}$ is a beta r.v. with parameters p and $(n-1)p$. Going back to equation (8) in Chapter 1, it follows that UMVUE of $R(t \mid p, \sigma)$ is given by $\tilde{R}(t)$, where

$$\tilde{R}(t) = \frac{1}{B\{p, (n-1)p\}} \int_{t/n\bar{x}}^{1} y^{p-1}(1-y)^{(n-1)p-1} dy$$

$$= 1 - I_{t/n\bar{x}}\{p, (n-1)p\} \quad \text{if } t < n\bar{x}$$

$$= 0 \quad \text{otherwise}$$

Here, $I_u(a, b)$ denotes the incomplete beta function $\dfrac{1}{B(a, b)} \displaystyle\int_0^u y^{a-1}(1-y)^{b-1}dy$.
We can thus obtain $\tilde{R}(t)$ by using the tables of incomplete beta functions (K. Pearson, 1968) or by numerical integration. The problem of obtaining UMVUE of $R(t \mid p, \sigma)$ for the gamma distribution when p and σ are both unknown, appears to be an open problem.

Next, we consider the case of Weibull distribution with both parameters p and σ unknown (or equivalently both parameters b and c unknown). The reliability function here is given by $R(t) = \exp\left\{-\dfrac{t^p}{\sigma}\right\} = \exp\left\{-\left(\dfrac{t}{b}\right)^c\right\}$.

The MLE of $\hat{R}(t)$ is then given by

$$\hat{R}(t) = \exp\left\{-\frac{t^{\hat{p}}}{\hat{\sigma}}\right\} = \exp\left\{-\left(\frac{t}{\hat{b}}\right)^{\hat{c}}\right\}$$

where $(\hat{p}, \hat{\sigma})$ are MLE of (p, σ) and $(\hat{b}, \hat{c})$ are MLE of (b, c).

For the data in Example 2.2 we have the following estimates for various values of t:

t	$\hat{R}(t)$	True value
1	0.7440	0.7788
2	0.2967	0.3679
3	0.0629	0.1054
4	0.0070	0.0183

For the case when p is known (or equivalently c known) the MLE of $R(t) = \exp\left(-\dfrac{t^p}{\sigma}\right) = \exp\left[-\left(\dfrac{t}{b}\right)^c\right]$ is

$$\hat{R}(t) = \exp\left\{-\frac{t^p}{\hat{\sigma}}\right\} = \exp\left\{-\left(\frac{t}{\hat{b}}\right)^c\right\}$$

where $\hat{\sigma} = \dfrac{\sum\limits_{i=1}^n x_i^p}{n}$ and $\hat{b} = \left[\dfrac{\sum\limits_{i=1}^n x_i^c}{n}\right]^{1/c}$. As in the case of gamma distribution with known p, it is possible to obtain UMVUE of $R(t)$ by considering the complete sufficient statistics $\hat{\sigma}$ for σ. Note that $Y = X^p$ is distributed as exponential density with parameter σ when X is Weibull with parameters p and σ and $\bar{Y} = \dfrac{1}{n}\sum\limits_{i=1}^n y_i = \hat{\sigma}$ is complete sufficient for σ since p is known. Further, $P\{X > t\} = P\{Y > t^p\}$. Thus, the reliability function $R(t)$ for Weibull with known p becomes the reliability function $R(t^p)$ for the exponential density. It, therefore, immediately follows that UMVUE of $R(t)$ say $\tilde{R}(t)$ is given by UMVUE of $R(t^p)$ in the exponential case. Thus from the results of Chapter 1 we have

$$\tilde{R}(t) = \left\{ 1 - \frac{t^p}{\sum_{i=1}^{n} x_i^p} \right\}^{(n-1)} \quad \text{if } t^p < \sum_{i=1}^{n} x_i^p$$

$$= 0 \qquad\qquad\qquad \text{otherwise}$$

EXERCISES

2.1 Prove that if a random variable U has a p.d.f. $f(u)$ over $(-\infty, \infty)$, then $V = [1 - F(U)]$, where F, the d.f. of U is uniformly distributed over $(0, 1)$ and $W = \log [- \log V]$ is distributed with p.d.f. $h(w) = \exp \{w - e^{-w}\}$, $-\infty < w < \infty$.

2.2 The following is the life distribution (in units of 100 hours) of 25 electron tubes:

0.1415	0.5937	2.3467	3.1356	3.5681
0.3484	1.1045	2.4651	3.2259	3.7287
0.3994	1.7323	2.6155	3.4177	9.2817
0.4174	1.8348	2.7425	3.5351	9.3208
0.5625	1.8474	2.9487	3.5680	17.2271

Assuming the distribution is given by a gamma p.d.f.

$$f(x \mid \sigma, p) = \frac{1}{\Gamma(p)\sigma} \left(\frac{x}{\sigma}\right)^{p-1} \exp\left(-\frac{x}{\sigma}\right), \quad x, p, \sigma > 0$$

compute the
(i) maximum likelihood, and
(ii) moment estimates of p and σ.

2.3 Compute the maximum likelihood estimates of $R(t)$ at $t = 2(2)10$ for data in 2.2.

2.4 Consider the following life distribution of 10 units (in hours):

30, 65, 125, 150, 155, 165, 180, 185, 192, 195.

Assuming the underlying distribution

$$f(x \mid \sigma) = \frac{x}{\sigma^2} \exp\left(-\frac{x}{\sigma}\right), \quad x, \sigma > 0$$

compute the maximum likelihood and minimum variance unbiased estimates of $R(t)$ at $t = 50(50)200$.

2.5 Consider the failure time distribution (in hours) of 15 items:

33.85	155.08	259.06
70.00	171.75	281.77
87.96	183.15	298.57
97.35	189.17	306.56
132.83	232.62	365.87

If the distribution is given by

$$f(x \mid \sigma) = \frac{2x}{\sigma} \exp\left(-\frac{x^2}{\sigma}\right), \quad x, \sigma > 0$$

compute the maximum likelihood and minimum variance unbiased estimates of $R(t)$ at $t = 200(50)400$.

2.6 A random sample of 30 observations was generated from the Weibull p.d.f.

$$f(x \mid \sigma, p) = \frac{p}{\sigma} x^{p-1} \exp\left(-\frac{x^p}{\sigma}\right), \quad x, p, \sigma > 0$$

(i) obtain (a) moment, (b) maximum likelihood, and (c) Menon's estimates of (p, σ).
(ii) estimate the coefficient of correlation between the maximum likelihood estimates of p and σ.
(iii) obtain the MLE of $R(t \mid p, \sigma)$ at $t = 0.5(.5)3.0$.

SAMPLE

2.124345	1.283969	2.050806	1.986794	1.990020
2.992270	0.891173	1.163034	1.026030	1.917587
1.001437	2.185564	1.316525	1.312445	1.991337
2.126520	1.864766	2.045745	1.549771	1.223020
0.753987	2.110567	1.213129	1.230608	2.169475
1.641153	3.508177	1.063310	1.014592	0.762945

2.7 Based on the first 20 ordered observations in the data of 2.6.
(i) compute the maximum likelihood estimates of p and σ and estimate the coefficient of correlation between the estimates,
(ii) obtain the MLE of $R(t \mid p, \sigma)$ at $t = 0.5(.5)3.0$.

2.8 The following is a random sample of 25 observations from the Weibull p.d.f

$$f(x \mid b, c) = \left(\frac{c}{b}\right)\left(\frac{x}{b}\right)^{c-1} \exp\left\{-\left(\frac{x}{b}\right)^c\right\}, \quad x, b, c > 0$$

SAMPLE

0.99995	1.83153	2.59064	3.65876	0.58544
3.06563	0.87957	1.55085	2.03406	2.40778
1.89170	2.81774	0.33847	1.03291	1.34652
0.97347	1.71750	2.32616	2.92412	3.55062
2.98572	0.70000	1.32834	1.71278	1.95078

Given $c = 2$ and the first ten ordered observations, compute the following:
(i) the MLE of b and an estimate of its standard error;
(ii) an unbiased estimate of b and an estimate of its standard error; and
(iii) the efficiency of the unbiased estimates of b based on the first ten ordered observations with respect to its maximum likelihood estimator based on the total observations.

2.9 Refer to the data in the Example 2.5. The last five ordered observations are 3.69354, 3.75785, 3.77393, 3.77450, 3.86484. Using the sample of twenty observations from the Weibull p.d.f. in (17) obtain White's estimators and the MLE of b and c. (Given the corresponding $X = 0.37756, 0.53042, 0.70223, 0.91195, 1.22321$.)

2.10 Estimate the reliability function at $t = 1, 2, 3, 4$ using the estimates obtained in (i) Exercise 2.9 above and (ii) Example 2.5.

2.11 50 units were placed on test and the test was terminated after the first r units failed. The rth unit failed at 100 hours from the starting time. The following life distribution was assumed:

$$f(x \mid \sigma) = \frac{2x}{\sigma} \exp\left(-\frac{x^2}{\sigma}\right), \quad x, \sigma > 0$$

and the MLE of σ was computed to be 5,000. Obtain the MLE of $R(t)$ at $t = 100$ and obtain the number of items that failed before 100 hours, assuming that the records were lost.

CHAPTER 3

Normal and Related Distributions

In this chapter we will consider the normal and log-normal as models for failure time distributions. We will consider the normal distribution first. The PDF of the normal distribution with location parameter μ and scale parameter σ is given by $f(x \mid \mu, \sigma) = \dfrac{1}{\sqrt{2\pi\sigma^2}} \left\{ \exp -\dfrac{1}{2}\left(\dfrac{x-\mu}{\sigma}\right)^2 \right\}$ $-\infty < x < \infty$, $-\infty < \mu < \infty$, $\sigma > 0$ which we will generally denote by $\dfrac{1}{\sigma}\phi\left(\dfrac{x-\mu}{\sigma}\right)$. The corresponding d.f. $F(x, \mu, \sigma)$ is given by

$$\int_{-\infty}^{x} \frac{1}{\sigma}\phi\left(\frac{u-\mu}{\sigma}\right) du = \Phi\left(\frac{x-\mu}{\sigma}\right)$$

There is no closed form for $\Phi\{(x-\mu)/\sigma\}$ available although both $\phi(x)$ and $\Phi(x)$ have been well tabulated in Pearson and Hartley (1954), Abramowitz and Stegan (1964). The parameters μ and σ also represent the mean and the standard deviation respectively, and σ^2 is the variance of the distribution.

The normal distribution plays a very important role in statistical theory as well as methods. The names of the great mathematicians such as Gauss, Laplace, Légendre and others are associated with the discovery and use of this distribution as a model for distribution of errors of measurement. Indeed, originally and perhaps the practice still continues, the p.d.f. of the normal distribution was called as the error function. The normal distribution also arises as a limiting form of various other distributions. Thus, for example, one can show that the gamma distribution considered in the previous chapter can be approximated by a normal distribution with mean $p\sigma$ and variance $p\sigma^2$ for large values of p. More generally, if $(X_1, X_2, \dots, X_n)$ are independent identically distributed r.v. with mean μ and variance σ^2, then for large n the sample mean $(1/n)\Sigma x_i = \bar{x}$ is approximately normal with mean μ and variance σ^2/n.

In the context of life testing and reliability problems, Davis (1952) has shown that the normal distributions give quite a good fit for the failure time data that he obtained on the failure times for 417 incandescent lamps. Bazovsky (1961) also discusses the use of normal distribution in life testing and reliability problems. For normal distribution we cannot obtain the instantaneous failure rate $\mu(t)$ in a closed form, although it can be shown that $\mu(t)$ is an increasing function of t which indicates the 'aging effect'.

Estimation of Parameters

Let $(x_1, x_2, \ldots, x_n)$ be a random sample of size n from normal distribution with mean μ and variance σ^2, denoted as $N(\mu, \sigma^2)$. Let $\bar{x} = \dfrac{1}{n} \sum_{i=1}^{n} x_i$ denote the sample mean and $s^2 = \sum_{i=1}^{n} (x_i - \bar{x})^2$ denote the sums of squares of deviations. Then, as is well known, $\bar{x}$ is itself $N\left(\mu, \dfrac{\sigma^2}{n}\right)$ and s^2/σ^2 is distributed as chi-squared with $(n - 1)$ degrees of freedom. Further, $\bar{x}$ and s^2 are stochastically independent. Thus, the p.d.f. of $v = s^2$ is given by

$$f(v \mid \sigma^2) = \frac{1}{\Gamma\left(\dfrac{n-1}{2}\right) 2^{n/2} \sigma^n} \, v^{((n-1)/2)-1} \exp\left(-\frac{v}{2\sigma^2}\right), \ v \geqslant 0, \ \sigma > 0$$

Thus, s^2 is a gamma variate with parameters $p = (n - 1)/2$ and the scale parameter $\sigma' = 1/2\sigma^2$. The likelihood of the sample is given by

$$L = \left(\frac{1}{\sqrt{2\pi}}\right)^n \frac{1}{\sigma^n} \exp\left\{-\sum_{i=1}^{u} \frac{(x_i - \mu)^2}{2\sigma^2}\right\}$$

and one can easily show that the MLE of μ and σ^2 are given by $\hat{\mu} = \bar{x}$ and $\hat{\sigma}^2 = s^2/n$. By using the factorizability criteria, one can show that $(\bar{x}, s^2)$ are jointly sufficient for (μ, σ^2). For proofs of these results, if the reader has not already come across them, we refer to the texts such as Larson (1969), Fraser (1958), among several others. Lehman and Schefé (1955) have shown that $(\bar{x}, s^2)$ are jointly complete for (μ, σ^2) and we can thus conclude that $\bar{x}$ and $s^2/(n - 1)$ (rather than s^2/n) are UMVUE of μ and σ^2 respectively.

Next, we will consider the case of failure censored samples. Here the data would be the r failure times arranged in order of magnitude $(x_1 < x_2 < \ldots < x_r)$ and the fact that $(n - r)$ items survived beyond x_r. We will obtain the MLE of μ and σ^2 in this case (Gupta, 1952). The likelihood of the sample is now given by

$$L \propto \prod_{i=1}^{n} \frac{1}{\sigma} \phi\left(\frac{x_i - \mu}{\sigma}\right) \left[1 - \Phi\left(\frac{x_r - \mu}{\sigma}\right)\right]^{n-r}$$

which immediately leads to

$$\log L = C - r \log \sigma - \frac{1}{2\sigma^2} \sum_{i=1}^{r} (x_i - \mu)^2 + (n - r) \log\left[1 - \Phi(\xi)\right]$$

where $\xi = \dfrac{x_r - \mu}{\sigma}$. The likelihood equations are given by

$$\frac{\partial \log L}{\partial \mu} = \frac{1}{\sigma^2} \sum_{i=1}^{r} (x_i - \mu) + \frac{(n - r)\phi(\xi)}{[1 - \Phi(\xi)]} \frac{1}{\sigma} = 0$$

$$\frac{\partial \log L}{\partial \sigma} = -\frac{r}{\sigma} + \frac{1}{\sigma^3} \sum_{i=1}^{r} (x_i - \mu)^2 + \frac{(n-r)\phi(\xi)}{[1 - \Phi(\xi)]} \frac{\xi}{\sigma} = 0$$

Let $\bar{x} = \frac{1}{r} \sum_{i=1}^{r} x_i$ and $d = x_r - \bar{x}$ and $s^2 = \frac{1}{r} \sum_{i=1}^{r} (x_i - \bar{x})^2$. Then the likelihood equations immediately lead to

$$\frac{\bar{x} - \mu}{\sigma} = -\frac{(n-r)\phi(\xi)}{r[1 - \Phi(\xi)]} \tag{1}$$

$$\frac{1}{\sigma^2} \sum_{i=1}^{r} (x_i - \mu)^2 = r - \frac{\xi(n-r)\phi(\xi)}{[1 - \Phi(\xi)]} \tag{2}$$

The above equations lead to

$$s^2 = \sigma^2 + (\bar{x} - \mu)d$$

or

$$\mu = (\sigma^2 - s^2)/d + \bar{x} \tag{3}$$

Let $p = r/n$, then from (1) we have

$$-\left(\frac{1}{p} - 1\right) \frac{\phi(\xi)}{1 - \Phi(\xi)} = \frac{\bar{x} - x_r + x_r - \mu}{\sigma} = -\frac{d}{\sigma} + \xi$$

or

$$\xi + \left(\frac{1}{p} - 1\right) \frac{\phi(\xi)}{1 - \Phi(\xi)} = \frac{d}{\sigma} \tag{4}$$

Denoting the LHS of (4) by z we immediately obtain $\sigma = d/z$. Now, $d = (x_r - \bar{x})$ can be written as $\xi\sigma - (\bar{x} - \mu)$ which in view of (3) leads to the following equations:

$$d = \xi\sigma + \frac{(\sigma^2 - s^2)}{d}$$

$$d^2 + s^2 = \xi\sigma d + \sigma^2$$

$$= \sigma^2\left(1 + \frac{\xi d}{\sigma}\right)$$

$$= \frac{d^2}{z^2}(1 + \xi z)$$

or

$$s^2 = \frac{d^2}{z^2}(1 + \xi z - z^2)$$

Thus,

$$\frac{d^2 + s^2}{s^2} = \frac{1 + \xi z}{1 + \xi z - z^2} \tag{5}$$

Now consider a linear transformation on observations, say $y = cx + d$ where $c > 0$. Then it is quite clear that $x_1 < x_2 < \ldots < x_r$ goes into $y_1 < y_2 < \ldots < y_r$ and y is $N(c\mu + d, c^2\sigma^2)$ which we denote by $N(\mu', \sigma'^2)$. Now $y_i - \bar{y} = c(x_i - \bar{x})$ for each $i = 1, 2, \ldots, r$ and, therefore, the quantity $\psi \equiv (d^2 + s^2)/d^2$ remains unchanged whether we use x observations or transformed y observations. Similarly, the quantity $\xi = (y_r - \mu')/\sigma' = (x_r - \mu)/\sigma$ remains unchanged and therefore z also remains unchanged since $p = r/n$ is fixed.

Thus, the equation (5) uniquely determines z for any combination of parametric values (μ, σ). Gupta (1952) has given the values of z for given $\psi = 0.05(.05)0.95$ and $p = 0.1(.1)1.0$ [see Table 3.2]. Once the value of $z = z_0$ is determined from this table, we can obtain the MLE of (μ, σ) quite easily and these are given by

$$\hat{\sigma} = \frac{d}{z_0} = \frac{x_r - \bar{x}}{z_0}$$

and

$$\hat{\mu} = \frac{\hat{\sigma}^2 - s^2}{d} + \bar{x}$$

Similar techniques can be applied to time censored samples where the experiment is terminated after fixed time $T = t_0$. The data here consists of individual life times of say r failures, $x_1 < x_2 < \ldots < x_r$, and the fact that $(n - r)$ items did not fail before t_0. The only change required in the above is to define $\xi = (t_0 - \mu)/\sigma$ and $d = t_0 - \bar{x}$.

The exact joint distribution of $(\hat{\mu}, \hat{\sigma})$ is very difficult to obtain. Gupta (1952) obtained asymptotic (large samples) expressions for $E(\hat{\mu})$, Var $(\hat{\mu})$, $E(\hat{\sigma})$, Var $(\hat{\sigma})$ and Cov $(\hat{\mu}, \hat{\sigma})$. The MLE $\hat{\mu}$ and $\hat{\sigma}$ are biased for μ and σ respectively although the bias tends to zero as $n \to \infty$.

Another method used quite often to estimate μ and σ in failure (or time-censored) samples is to use linear functions of order statistics. Considerable work has been done in this area and we refer to papers by Gupta (1952), Sarhan and Greenberg (1962) and Plackett (1959) among several others. Let $(X_{1n}, X_{2n}, \ldots, X_{rn})$ denote the first r order statistics in a sample of size n from $N(\mu, \sigma^2)$. Then we construct estimators of the type

$$\tilde{\mu} = \sum_{i=1}^{r} \beta_{in} x_{in}$$

$$\tilde{\sigma} = \sum_{i=1}^{r} \gamma_{in} x_{in}$$

where the coefficients $(\beta_{1n}, \ldots, \beta_{rn})$ and $(\gamma_{1n}, \gamma_{2n}, \ldots, \gamma_{rn})$ are chosen appropriately. To make the notation less cumbersome we drop the subscript n from the above notation. Let $Y_i = (X_i - \mu)/\sigma$, then Y_i is ith order statistics from $N(0, 1)$ and $E(Y_i) = \lambda_i$ and Cov $(Y_i, Y_j) = v_{ij}$. Thus, $E(X_i) = \mu + \lambda_i \sigma$ and Cov $(X_i, X_j) = \sigma^2 v_{ij}$. Now we select the coefficients $(\beta_1, \ldots, \beta_r)$ so that $E(\tilde{\mu}) = \mu$ and Var $(\tilde{\mu})$ is as small as possible. Thus, $\tilde{\mu}$ has minimum variance within the class of unbiased estimators of μ, of the type $\Sigma l_i x_i$. Similarly, the coefficients $(\gamma_1, \gamma_2, \ldots, \gamma_r)$ are chosen so that $E(\tilde{\sigma}) = \sigma$ and Var $(\tilde{\sigma})$ is minimum.

The coefficients $(\beta_1, \ldots, \beta_r)$ and $(\gamma_1, \gamma_2, \ldots \gamma_r)$ would be functions of $(\lambda_1, \ldots \lambda_r)$ and v_{ij}, $i = 1, 2, \ldots r$, $j = 1, 2, \ldots r$. For $s \leqslant r < n \leqslant 10$ Gupta (1952) has tabulated these coefficients. For $n = 2(1), 20$, Teichrow (1962) gives the values of λ_i and v_{ij} for $2 \leqslant r \leqslant 10$. For the values of r between 10 and 20, we can use the following relations:

$$\lambda_i = -\lambda_{(n-i+1)}, \quad \text{i.e.,} \quad E(X_i) = -E(X_{(n-i+1)})$$

$$v_{ij} = v_{(n-i+1)(n-j+1)} \qquad \text{Cov } (X_i, X_j) = \text{Cov } (X_{(n-i+1)}, X_{(n-j+1)})$$

Sarhan and Greenberg (1962) have published tables of coefficients for obtaining the minimum variance unbiased estimators of μ and σ of the type

$$\mu^* = \sum_{i=r_1+1}^{n-r_2} a_{1i} x_i \tag{6}$$

$$\sigma^* = \sum_{i=r_1+1}^{n-r_2} a_{2i} x_i \tag{7}$$

These are available for $n = 2(1), 20, r_1 = 0(1)8$, and $r_2 = 0(1), 14$. For the failure (or time) censored samples $r_1 = 0$ and $r_2 = n - r$ where r is the preassigned (or random) number of failures. For the values of $n = 20(1)$, 50 we can obtain the values of x_i from Fisher and Yates (1957) but the constants v_{ij}'s are not readily available and are quite difficult to compute. In this case we can use alternative estimators suggested by Gupta (1952) and Plackett (1959) which depend only on $(\lambda_1, \ldots, \lambda_n)$. These alternative estimators would have, as to be expected, larger variance than the estimators obtained earlier.

Gupta (1952) suggested the following estimators:

$$\tilde{\mu}^* = \sum_{i=1}^{r} b_i x_i \tag{8}$$

$$\tilde{\sigma}^* = \sum_{i=1}^{r} c_i x_i \tag{9}$$

where

$$b_i = \frac{1}{r} - \frac{\lambda_r (\lambda_i - \bar{\lambda}_r)}{\sum_{i=1}^{r} (\lambda_i - \bar{\lambda}_r)^2}$$

$$c_i = \frac{\lambda_i - \bar{\lambda}_r}{\sum_{i=1}^{r} (\lambda_i - \bar{\lambda}_r)^2}$$

and

$$\bar{\lambda}_r = \frac{1}{r} \sum_{i=1}^{r} \lambda_i.$$

Plackett (1959) suggested the following estimators:

$$\tilde{\mu}_p = \begin{vmatrix} \sum_{1}^{r} g_{1i} x_i & \sum_{i=1}^{r} g_{1i} \lambda_i \\ \sum_{1}^{r} g_{2i} x_i & \sum_{i=1}^{r} g_{2i} \lambda_i \end{vmatrix} \div \Delta \tag{10}$$

$$\tilde{\sigma}_p = \begin{vmatrix} \sum_{i=1}^{r} g_{1i} & \sum_{i=1}^{r} g_{1i} x_i \\ \sum_{i=1}^{r} g_{2i} & \sum_{i=1}^{r} g_{2i} x_i \end{vmatrix} \div \Delta \tag{11}$$

where

$$\Delta = \begin{vmatrix} \sum_{i=1}^{r} g_{1i} & \sum_{i=1}^{r} g_{1i}\lambda_i \\ \sum_{i=1}^{r} g_{2i} & \sum_{i=1}^{r} g_{2i}\lambda_i \end{vmatrix}$$

The constants appearing in the above formula are defined below:

$$g_{11} = \frac{\phi^2(\lambda_1)}{\Phi(\lambda_1)} + \lambda_1\phi(\lambda_1) + \frac{1}{2(n+1)}$$

$$g_{1i} = \frac{1}{n+1}, \quad i = 2, 3, \ldots, r-1$$

$$g_{1r} = \frac{\phi^2(\lambda_r)}{1 - \Phi(\lambda_r)} - \lambda_r\phi(\lambda_r) + \frac{1}{2(n+1)}$$

$$g_{21} = \frac{\lambda_1\phi^2(\lambda_1)}{\Phi(\lambda_1)} + \lambda_1^2\phi(\lambda_1) - \phi(\lambda_1) + \frac{\lambda_1}{n+1}$$

$$g_{2i} = \frac{2\lambda_i}{n+1}, \quad i = 2, 3, \ldots, r-1$$

$$g_{2r} = \frac{\lambda_r\phi^2(\lambda_r)}{1 - \Phi(\lambda_r)} - \lambda_r^2\phi^2(\lambda_r) + \phi(\lambda_r) + \frac{\lambda_r}{n+1}$$

EXAMPLE 3.1   12 units were placed on test and the experiment was terminated after 9 failures. The failures occurred at 980, 1030, 1120, 1260, 1320, 1390, 1408, 1450, 1485 hours. Assuming that the failure time distribution is $N(\mu, \sigma^2)$ we would estimate μ and σ using different methods discussed above.

First consider the MLE. Here $n = 12$, $r = 9$, $p = 0.75$, $x_r = 1485$, $\sum_{i=1}^{r} x_i = 11443$, $\bar{x} = 1271.4$, $s^2 = 34856.3$. This immediately leads to $d = x_r - \bar{x} = 213.6$, $\psi \equiv s^2/(s^2 + d^2) = 0.4331$. Using Table 3.2, by interpolation we obtain $z_0 = 0.9196$ and, therefore, the MLE are given by $\hat{\mu} = 1361$ and $\hat{\sigma} = 232$. By using Gupta's results we can obtain estimated standard errors of $\hat{\mu}$ and $\hat{\sigma}$ and these are given by SE $(\hat{\mu}) = 70$ and SE $(\hat{\sigma}) = 58$.

Next, suppose we want to obtain minimum variance unbiased estimators μ^* and σ^* as given in (6) and (7). We can directly use the tables 10c.1 and from Sarhan and Greenberg (1962) with $n = 12$, $r_1 = 0$, $r_2 = 3$ which lead to $\mu^* = 1367$ and $\sigma^* = 238$. The estimates of the standard errors of μ^* and σ^* can be obtained from the Table 10c.2 from Sarhan and Greenberg (1962) and these are given by SE $(\mu^*) = 75$ and SE $(\sigma^*) = 67$.

For the sake of comparison, we work out alternative estimators due to Gupta (1952) and Plackett (1959) as given in (8), (9) and (10), (11) respectively. First, consider estimators due to Gupta. Using Fisher and Yates (1957) Table 20, page 86, for $n = 12$, we have the values of λ_i, $i = 1, 2$,

..., 9 are given by -1.63, -1.12, -0.79, -0.54, -0.31, -0.10, 0.10, 0.31, 0.54. $\bar{\lambda}_r = -0.3933$, $\sum_1^r (\lambda_1 - \bar{\lambda}_r)^2 = 3.939$. The various values of b_i and c_i are as given below:

b_i	x_i	c_i
-0.0123	980	-0.3140
0.0387	1030	-0.1846
0.0715	1120	-0.1008
0.0964	1260	-0.0373
0.1195	1320	0.0211
0.1403	1390	0.0744
0.1603	1408	0.1252
0.1812	1450	0.1785
0.2043	1485	0.2369

This leads to $\tilde{\mu}^* = 1376$ and $\tilde{\sigma}^* = 260$. To obtain the estimates of SE $(\tilde{\mu}^*)$ and SE $(\tilde{\sigma}^*)$ we first obtain Var $(\tilde{\mu}^*)$ and Var $(\tilde{\sigma}^*)$. Since the estimators are linear functions of the order statistics these are given by

$$\text{Var} (\hat{\mu}^*) = \sigma^2 b'Vb \quad \text{and} \quad \text{Var} (\tilde{\sigma}^*) = \sigma^2 c'Vc$$

where V is the variance-covariance matrix $\|v_{ij}\|$, $\quad i = 1, 2, \ldots, 9$

$$j = 1, 2, \ldots, 9$$

and b and c are the column vectors of b_i and c_i given above.

$$V = \begin{bmatrix} 0.3237 & 0.1603 & 0.1090 & 0.0831 & 0.0672 & 0.0559 & 0.0477 & 0.0409 & 0.0354 \\ & 0.1973 & 0.1350 & 0.1033 & 0.0836 & 0.0697 & 0.0595 & 0.0511 & 0.0442 \\ & & 0.1581 & 0.1212 & 0.0983 & 0.0823 & 0.0701 & 0.0604 & 0.0523 \\ & & & 0.1398 & 0.1136 & 0.0951 & 0.0813 & 0.0701 & 0.0607 \\ & & & & 0.1306 & 0.1096 & 0.0937 & 0.0809 & 0.0701 \\ & & & & & 0.1267 & 0.1084 & 0.0937 & 0.0813 \\ & & & & & & 0.1267 & 0.1096 & 0.0951 \\ & & & & & & & 0.1306 & 0.1136 \\ & & & & & & & & 0.1398 \end{bmatrix}$$

$$b'Vb = 0.0956537$$
$$c'Vc = 0.0784322$$

Hence estimated SE $(\mu^*) = 80$ and SE $(\sigma^*) = 73$.

Next, we consider Plackett's estimators (10) and (11). We note that $\lambda_1 = -1.63$, $\phi(\lambda_1) = 0.106$, $\Phi(\lambda_1) = 0.052$, $\lambda_9 = 0.54$, $\phi(\lambda_9) = 0.345$, and $1 - \Phi(\lambda_9) = 0.295$. The various g_{1i} and g_{2i} values along with x_i and λ_i are given below.

g_{1i}	x_i	λ_i	g_{2i}
0.0817584	980	−1.63	−0.3019590
0.0769231	1030	−1.12	−0.1723077
0.0769231	1120	−0.79	−0.1215385
0.0769231	1260	−0.54	−0.0830769
0.0769231	1320	−0.31	−0.0476923
0.0769231	1390	−0.10	−0.0153846
0.0769231	1408	0.10	0.0153846
0.0769231	1450	0.31	0.0476923
0.2556361	1485	0.54	0.5038128

These lead to

$$\sum_1^9 g_{1i} = 0.8758562 \qquad \sum_1^9 g_{1i}x_i = 1150.3582836 \qquad \sum_1^9 g_{1i}\lambda_i = -0.1836844$$

$$\sum_1^9 g_{2i} = -0.1750693 \qquad \sum_1^9 g_{2i}x_i = 40.4421648 \qquad \sum_1^9 g_{2i}\lambda_i = 1.1307598$$

Applying formula (10) and (11) we obtain $\tilde{\mu}_p = 1356$ and $\tilde{\sigma}_p = 247$. Plackett (1959) has also given the following formula for the variances of $\tilde{\mu}_p$ and $\tilde{\sigma}_p$.

$$\text{Var}(\tilde{\mu}_p) = \sigma^2 \left\{ \sum_{i=1}^r g_{2i}\lambda_i / n\Delta \right\}$$

and

$$\text{Var}(\tilde{\sigma}_p) = \sigma^2 \left\{ \sum_{i=1}^r g_{1i}\lambda_i / n\Delta \right\}$$

The estimates of $\text{SE}(\tilde{\mu}_p) = 77$ and $\text{SE}(\tilde{\sigma}_p) = 68$. Table 3.1 summarizes the various estimates of μ and σ obtained by different methods discussed above and the estimates of their standard errors.

Table 3.1   A Comparison of the Estimators of μ and σ

	μ_E	σ_E	*Est* SE(μ_E)	*Est* SE(σ_E)
Gupta (MLE)	1361	232	70	58
Sarhan-Greenberg (Unbiased)	1367	248	75	67
Gupta (Alternative)	1376	260	80	73
Plackett	1365	247	77	68

Log-Normal Distribution

In this section we consider the log-normal distribution. Very simply, the log-normal distribution arises if we assume that a non-negative r.v. X is such that $U = \log X$ is $N(\mu, \sigma^2)$. Thus, instead of assuming that the failure times are independent normals, we now assume $\log X_i$ are independent $N(\mu, \sigma^2)$. The log-normal distribution arises in various different contexts such as in physics (distribution of particles due to pulverization); economics (income distributions); biology (growth of organism), etc.

Aitchison and Brown (1957) have given a very comprehensive treatment of log-normal distribution. We refer to the papers by Epstein (1947), Brownlee (1949), Delaporte (1950), Moroney (1951) which describe various applications of log-normal distribution to physical and industrial processes, textile research and quality control. In the context of life testing and reliability problems, the log-normal distribution answers a criticism sometimes raised against the use of normal distribution (ranging from $-\infty$ to ∞) as a model for failure time distribution which must range over 0 to ∞. We remark here that the above criticism could be levelled against every case in which an essentially non-negative quantity is assumed to be $N(\mu, \sigma^2)$. However, it is well-known that the normal distribution represents very adequately variables such as height, shoe sizes of men, lengths of beans, etc. This is mainly due to the fact that the probability content outside $(\mu - 5\sigma, \mu + 5\sigma)$ is of the order 10^{-7}, that outside $(\mu - 10\sigma, \mu + 10\sigma)$ is of the order 10^{-9}.

The p.d.f. of the log-normal distribution is given by

$$f(x, \mu, \sigma) = \frac{1}{\sqrt{2\pi}\sigma x} \exp\left\{-\frac{1}{2\sigma^2} (\log x - \mu)^2\right\}, \, x > 0, \, -\infty < \mu < \infty, \, \sigma > 0$$

$$= \frac{1}{\sigma x} \phi\left(\frac{\log x - \mu}{\sigma}\right) \tag{12}$$

The corresponding d.f. $F(t)$ is given by

$$F(t, \mu, \sigma) = \int_0^t \frac{1}{\sqrt{2\pi}\sigma x} \exp\left\{-\frac{1}{2\sigma^2} (\log x - \mu)^2\right\} dx$$

$$= \int_{-\infty}^{\log t} \frac{1}{\sqrt{2\pi}\sigma} \exp\left[-\frac{1}{2\sigma^2} (u - \mu)^2\right] du$$

$$= \Phi\left(\frac{\log t - \mu}{\sigma}\right) \tag{13}$$

The reliability function $R(t, \mu, \sigma) = 1 - F(t, \mu, \sigma) = 1 - \Phi\{(\log t - \mu)/\sigma\}$. Again, as in the normal case the instantaneous hazard rate $\mu(t)$ cannot be obtained in a closed form. However, one can prove that for log-normal distribution the hazard rate $\mu(t)$ can be shown to be a decreasing function of time t. Thus, the log-normal distribution would also serve as a good model when the failure rate is rather high initially and then decreases as time increases. For the normal distribution the opposite is true—namely, $\mu(t)$ is an increasing function of time t. Thus, the choice of log-normal distribution as a model for failure time distribution should be made rather carefully. We also note that while the normal distribution is symmetric about μ, the log-normal distribution is quite skewed and this aspect should also be taken into consideration.

Now consider the expected value of the log-normal distribution

$$\alpha = E(X) = \int_0^\infty \frac{1}{\sqrt{2\pi}\sigma x} \times \exp\left\{-\frac{1}{2\sigma^2} (\log x - \mu)^2\right\} dx$$

Making the transformation $u = \log x$ we have

$$\alpha = \int_{-\infty}^{\infty} \frac{1}{\sqrt{2\pi}\sigma} \exp\left\{-\frac{1}{2\sigma^2}(u-\mu)^2 + u\right\} du$$

$$= \int_{-\infty}^{\infty} \frac{1}{\sqrt{2\pi}\sigma} \exp\left\{-\frac{1}{2\sigma^2}[(u-\mu)^2 - 2\sigma^2(u-\mu) + \sigma^4 - 2\sigma^2\mu - \sigma^4]\right\} du$$

$$= \exp\left(\mu + \frac{\sigma^2}{2}\right)$$

Similarly, we can show that

$$E(X^2) = \exp(2\mu + 2\sigma^2)$$

Therefore, Var $(X) = \beta = \exp(2\mu + \sigma^2)\{\exp(\sigma^2) - 1\}$.

We will now consider the estimation of the parameters α and β. Let $u_i = \log x_i$, $i = 1, 2, \ldots, n$. Then as has been seen earlier, using the fact that $(u_1, u_2, \ldots, u_n)$ is a random sample from $N(\mu, \sigma^2)$ we can obtain the MLE of μ and σ^2 first. These are given by

$$\hat{\mu} = \frac{1}{n}\sum_{i=1}^{n} u_i = \bar{u} = \frac{1}{n}\sum_{i=1}^{n} \log x_i \tag{14}$$

and

$$\hat{\sigma}^2 = \frac{1}{n}\sum_{i=1}^{n}(u_i - \bar{u})^2 = \frac{s_u^2}{n} = \frac{1}{n}\left\{\sum_{i=1}^{n}\left(\log x_i - \frac{1}{n}\sum_{i=1}^{n}\log x_i\right)^2\right\} \tag{15}$$

We note that $\bar{u}$ and $s_u^2/(n-1)$ are UMVUE of μ and σ^2. The MLE of α and β are given by

$$\hat{\alpha} = \exp\left\{\hat{\mu} + \frac{\hat{\sigma}^2}{2}\right\} = \exp\left\{\bar{u} + \frac{s_u^2}{2n}\right\} \tag{16}$$

$$\hat{\beta} = \exp(2\hat{\mu} + \hat{\sigma}^2)\{\exp(\hat{\sigma}^2) - 1\}$$

$$= \exp\left\{2\bar{u} + \frac{s_u^2}{n}\right\}\left\{\exp\left(\frac{s_u^2}{n}\right) - 1\right\} \tag{17}$$

where $\bar{u}$ and s_u^2 are defined as above.

It can be shown that the bias of $\hat{\alpha}$ and $\hat{\beta} \to 0$ as $n \to \infty$ and the variances of $\hat{\alpha}$ and $\hat{\beta}$ are given by

$$V(\hat{\alpha}) = \exp\left(2\mu + \frac{\sigma^2}{n}\right)\left[\exp\left(\frac{\sigma^2}{n}\right)\left(1 - \frac{2\sigma^2}{n}\right)^{-(n-1)/2} - \left(1 - \frac{\sigma^2}{n}\right)^{-(n-1)}\right] \tag{18}$$

$$V(\hat{\beta}) = \exp\left(4\mu + \frac{4\sigma^2}{n}\right)\left[\exp\left(\frac{4\sigma^2}{n}\right)\left\{\left(1 - \frac{8\sigma^2}{n}\right)^{-(n-1)/2}\right.\right.$$

$$-2\left(1 - \frac{6\sigma^2}{n}\right)^{-(n-1)/2} + \left(1 - \frac{4\sigma^2}{n}\right)^{-(n-1)/2}\right\}$$

$$-\left\{\left(1 - \frac{4\sigma^2}{n}\right)^{-(n-1)} + \left(1 - \frac{2\sigma^2}{n}\right)^{-(n-1)}\right.$$

$$\left.\left. -2\left(1 - \frac{6\sigma^2}{n} + \frac{8\sigma^4}{n}\right)^{-(n-1)/2}\right\}\right] \tag{19}$$

The problem of obtaining UMVUE for α and β is rather difficult and will not be attemped here. Indeed, not much is known in this direction.

EXAMPLE 3.2 A random sample of 30 observations was generated from a log-normal distribution

$$g(x) = \frac{1}{\sqrt{2\pi}\sigma x} \exp\left\{-\frac{1}{2\sigma^2}(\log x - \mu)^2\right\}, 0 < x < \infty$$

with $\mu = 2$ and $\sigma = 1$.

SAMPLE X

6.760	2.593	15.800	27.221	11.728	21.824
24.434	4.389	7.456	10.268	1.255	12.391
1.489	4.522	2.401	4.063	8.516	13.383
3.414	1.872	4.179	9.718	12.910	1.508
3.287	6.841	30.631	3.337	2.266	17.340

Obtain the MLE of α and β and estimate the standard errors of these estimates. We have $\bar{u} = 1.8363$, $s_u^2 = 25.145$.

From (16) and (17), the MLE

$$\hat{\alpha} = \exp(2.2554) = 9.5391$$

$$\hat{\beta} = \exp(5.3490) - \exp(4.5108) = 119.4032$$

From (18) and (19),

estimated standard error of $\hat{\alpha} = 1.94$

estimated standard error of $\hat{\beta} = 113.02$

Parameters	True values	MLE	Estimated SE
α	12.18	9.54	1.94
β	255.02	119.40	113.02

For failure censored samples, we can obtain the MLE of α and β by using the normal theory developed in the previous section by using the transformed observations $u_i = \log x_i$, $i = 1, 2, \ldots, r$ where r is the number of failures. If we are dealing with the time censored sample, then the time of termination t_0 should also be transformed to $\log t_0$. Note that $x_i < t_0$ holds if, and only if, $\log x_i < \log t_0$ holds. Thus, the result of Gupta (1952) on censored samples from the normal distribution is applicable to the log-normal case with the above modifications. For a more complete treatment of MLE of parameters of the log-normal distribution from censored samples we refer to papers by Harter and Moore (1966) and Tiku (1968). Not much work has been done on the estimation of the mean and the standard deviation of the log-normal distribution by using functions of linear combinations of order statistics $x_{(1)} < x_{(2)} < x_{(3)} \ldots < x_{(r)}$.

As in the case of the exponential, gamma and Weibull distributions, the log-normal distribution can be generalized to take account of the fact that the item does not fail before time k. Consider a three parameter log-normal distribution. Let X be a standard normal variate and let the log-normal variate Y be connected with X by the relation

$$X = \frac{1}{\sigma} [\log (Y - k) - \mu]$$

The probability density function of Y is given by

$$g(y) = \frac{1}{\sqrt{2\pi}\sigma(y - k)} \exp \left[-\frac{1}{2\sigma^2} \{\log (y - k) - \mu\}^2 \right], \quad k < y < \infty$$

Suppose we have a random sample $(Y_1, Y_2, \ldots, Y_n)$ from a three-parameter log-normal distribution defined above and we wish to estimate (k, μ, σ^2). If k is known, by using the transformation $z = y - k$, we reduce the problem to the two-parameter log-normal distribution discussed earlier.

When k is unknown, it is quite tempting to take the MLE of k as $\hat{k} = y_{(1)}$. But in this case, the MLE of (μ, σ^2) are difficult to obtain and indeed $(\hat{\mu}, \hat{\sigma}^2)$ could be $(-\infty, \infty)$ (Hill, 1963). We, therefore, will consider the moment estimators of (k, μ, σ^2).

We evaluate the first three moments of the three-parameter log-normal distribution.

$$\mu_1' = \int_k^\infty yg(y)\, dy$$

$$= \frac{1}{\sqrt{2\pi}} \int_{-\infty}^\infty \{k + \exp(\mu + x\sigma)\} \exp\left(\frac{-x^2}{2}\right) dx$$

$$= k + \sqrt{\lambda} \exp(\mu)$$

The central rth moment

$$\mu_r = \int_k^\infty (y - \mu_1')^r g(y)\, dy$$

$$= \frac{1}{\sqrt{2\pi}} \int_{-\infty}^\infty \{\exp(x\sigma + \mu) - \sqrt{\lambda}\exp(\mu)\}^r \exp\left(\frac{-x^2}{2}\right) dx$$

$$= \left[\lambda^{r^2/2} - \binom{r}{1}\lambda^{\{(r-1)^2+1\}/2} + \binom{r}{2}\lambda^{\{(r-2)^2+2\}/2} - \binom{r}{3}\lambda^{\{(r-3)^2+3\}/2} \right.$$

$$\left. + \binom{r}{4}\lambda^{\{(r-4)^2+4\}/2} - \ldots + (-1)^r\lambda^{r/2} \right] \exp(\mu r)$$

Hence,

$$\mu_2 = \lambda(\lambda - 1)\exp(2\mu)$$

$$\mu_3 = \lambda^{3/2}(\lambda - 1)^2(\lambda + 2)\exp(3\mu)$$

$$\beta_1 = \frac{\mu_3^2}{\mu_2^3} = (\lambda - 1)(\lambda + 2)^2$$

Note that β_1 is a measure of skewness and is a function of $\lambda = \exp(\sigma^2)$. We now outline the procedure for obtaining the moment estimators:

(1) Compute (m_1', m_2', m_3'), the first three raw moments from the data, the sample central moments $m_2 = m_2' - m_1'^2$, $m_3 = m_3 - 3m_2'm_1' + 2m_1'^3$, and $b_1 = (m_3^2/m_2^3)$, the measure of skewness for the sample.

(2) Estimate $\widehat{\beta_1} = b_1$ and $\widehat{\lambda}$ by using $b_1 = \widehat{\beta_1} = (\widehat{\lambda} - 1)(\widehat{\lambda} + 2)^2$. This will estimate $\widehat{\sigma}^2 = \log \widehat{\lambda}$.

(3) Obtain $\widehat{\mu}$ by solving the equation $m_2 = \widehat{\lambda}(\widehat{\lambda} - 1) \exp(2\widehat{\mu})$.

(4) Estimate k by solving $m_1' = \overline{y} = \widehat{k} + \sqrt{\widehat{\lambda}} \exp(\widehat{\mu})$.
We illustrate this technique by the following example.

EXAMPLE 3.3 A random sample of size 50 was generated from the log-normal distribution with $k = 6$, $\mu = -1$, $\sigma = \frac{1}{2}$.

SAMPLE

7.224965	6.278713	7.247746	6.194886	6.296265
6.347706	6.570497	6.292591	6.273369	6.197176
6.219232	6.196571	6.456061	7.501055	6.401834
6.156997	6.396584	6.209719	6.294649	6.369959
6.467147	6.418496	6.920742	6.708300	6.515878
6.575296	6.217236	6.273452	6.212994	6.234104
6.656303	6.356572	6.315824	6.575912	6.333712
6.293339	6.378195	6.145265	6.125301	6.179804
6.264914	6.230859	6.256852	6.400565	6.423813
6.473993	6.245519	6.927077	6.381112	6.155843

$$m_1' = 6.415820, \quad m_2' = 41.247234, \quad m_3' = 265.768933$$

$$m_2 = 0.084488, \quad m_3 = 0.0499979, \quad \widehat{\beta_1} = 4.141730$$

$$\widehat{\lambda} = 1.365635$$

Parameters	True Values	Moment Estimates
μ	−1.0	−0.888
σ	0.5	0.558
k	6.0	5.935

Recently Giesbrecht and Kempthorne (1976) have obtained the MLE of (k, μ, σ^2) by grouping the data. This technique avoids the maximum likelihood estimation problem for unbounded likelihood and subsequent determination of maximum likelihood estimators for three-parameter log-normal distribution. For more details and further generalizations of the log-normal distribution we refer to Aitchison and Brown (1957) and B. M. Hill (1963).

Reliability Estimation

Let $(X_1, X_2, \ldots, X_n)$ be a random sample from a normal distribution with p.d.f.

$$f(x \mid \mu, \sigma) = \frac{1}{\sqrt{2\pi}\sigma} \exp\left\{-\frac{(x-\mu)^2}{2\sigma^2}\right\}, \quad -\infty < x < \infty,$$

$$-\infty < \mu < \infty, \sigma > 0$$

Consider the case when σ is known. $\bar{x}$ is the MLE of μ and the MLE of reliability function

$$R(t \mid \mu) = 1 - \Phi\left(\frac{t-\mu}{\sigma}\right)$$

is given by $R(t \mid \hat{\mu}) = \hat{R}(t)$ say, where

$$\hat{R}(t) = 1 - \Phi\left(\frac{t-\bar{x}}{\sigma}\right)$$

$\bar{X}$ being a complete sufficient statistic, the UMVUE of $R(t)$ is given by $\tilde{R}(t) = E[T(x) \mid \bar{x}]$, where $T(x)$ is any unbiased estimator of $R(t)$. Take $T(x_1) = 1$ if $x_1 \geqslant t$ and $= 0$ otherwise, we have

$$\tilde{R}(t) = P(X_1 \geqslant t \mid \bar{x}) \tag{20}$$

To obtain $\tilde{R}(t)$, we have to know the conditional distribution of X_1, given $\bar{x}$ and this we derive in the following way. We note that $(X_1, \bar{X})$ are both linear combinations of independent normal variables and as such the joint distribution of $(X_1, \bar{X})$ is bivariate normal with means $E(X_1) = \mu = E(\bar{X})$ and variances $\mathrm{Var}(X_1) = \sigma^2$, $\mathrm{Var}(\bar{X}) = \sigma^2/n$ and the $\mathrm{Cov}(X_1, \bar{X}) = \sigma^2/n$. Thus the joint p.d.f. of $(X_1, \bar{X})$ is given by

$$g(x_1, \bar{x}) = \frac{1}{\frac{2\pi\sigma^2}{\sqrt{n}}\sqrt{\left(1-\frac{1}{n}\right)}} \exp\left[-\frac{1}{2\left(1-\frac{1}{n}\right)}\left\{\left(\frac{x_1-\mu}{\sigma}\right)^2\right.\right.$$

$$\left.\left. + n\left(\frac{\bar{x}-\mu}{\sigma}\right)^2 - \frac{2(\bar{x}-\mu)(x_1-\mu)}{\sigma^2}\right\}\right]$$

$$= \frac{1}{\sqrt{\frac{2\pi\sigma^2(n-1)}{n}}} \exp\left[\frac{-(x_1-\bar{x})^2}{\frac{2\sigma^2}{n}(n-1)}\right] \frac{1}{\sqrt{\frac{2\pi\sigma^2}{n}}} \exp\left[-\frac{(\bar{x}-\mu)^2}{\frac{2\sigma^2}{n}}\right]$$

Therefore, the conditional p.d.f. of X_1 given $\bar{X} = \bar{x}$ is

$$h(x_1, \bar{x}) = \frac{1}{\sqrt{\frac{2\pi\sigma^2(n-1)}{n}}} \exp\left[-\frac{(x_1-\bar{x})^2}{\frac{2\sigma^2}{n}(n-1)}\right]$$

Thus given $\bar{X} = \bar{x}$, X_1 is normally distributed with mean $= \bar{x}$ and variance $= \sigma^2(1 - 1/n)$. From (20) we have

$$\tilde{R}(t) = P\left[\frac{X_1-\bar{x}}{\sqrt{\sigma^2\left(1-\frac{1}{n}\right)}} \geqslant \frac{t-\bar{x}}{\sqrt{\sigma^2\left(1-\frac{1}{n}\right)}}\right] = 1 - \Phi\left\{\frac{t-\bar{x}}{\sqrt{\sigma^2\left(1-\frac{1}{n}\right)}}\right\}$$

(Zacks and Even, 1966)

For large n, the asymptotic behaviour of $\hat{R}(t)$ and $\tilde{R}(t)$ is the same. Zacks and Even (1966) have tabulated the efficiencies of the two estimations for varying n and known σ.

Next, consider the case when both μ and σ are unknown. As noted earlier the MLE are given by $\hat{\mu} = \bar{x}$ and $\hat{\sigma}^2 = s^2/n$, where $s^2 = \sum_{i=1}^{n}(x_i - \bar{x})^2$. It follows then that the MLE of $R(t \mid \mu, \sigma)$ is given by

$$R(t \mid \hat{\mu}, \hat{\sigma}) = \hat{R}(t) = 1 - \Phi\left\{\frac{\sqrt{n}(t - \bar{x})}{s}\right\}$$

The problem of obtaining UMVUE of $R(t \mid \mu, \sigma)$ when both parameters are unknown is slightly difficult but the derivation is quite instructive. The UMVUE of $R(t \mid \mu, \sigma)$ is given by $P(X_1 \geq t \mid \bar{x}, s)$ and one should therefore try to obtain the conditional distribution of X_1 given $\bar{X} = \bar{x}$ and $S^2 = s^2$. This again is quite complex, so we proceed as follows.

$$P(X_1 \geq t \mid \bar{x}, s^2) = P\left\{\frac{X_1 - \bar{x}}{\sqrt{\frac{(n-1)s^2}{n}}} \geq \frac{t - \bar{x}}{\sqrt{\frac{(n-1)s^2}{n}}} \mid \bar{x}, s^2\right\}$$

Let us now denote by $U = \dfrac{X_1 - \bar{X}}{\sqrt{(n-1)s^2/n}}$, where $\bar{X} = \dfrac{1}{n}\sum_{i=1}^{n} X_i$ and $s^2 = \sum_{i=1}^{n}(X_i - \bar{X})^2$. Now the statistics U^2 has a distribution which is independent of μ and σ^2. This follows from the fact that U^2 remains the same under the transformation $Y_i = (X_i - \mu)/\sigma$, $i = 1, 2, \ldots, n$ and $Y_1, Y_2, \ldots, Y_n$ are independent $N(0, 1)$ r.v.s. Therefore by Basu's theorem (see e.g. Hogg and Craig, 1978), U is independent of $(\bar{x}, s^2)$. Thus,

$$P\{X_1 \geq t \mid \bar{x}, s^2\} = P\left\{U \geq \frac{t - \bar{x}}{\sqrt{\frac{(n-1)s^2}{n}}}\right\}$$

where $U^2 = \dfrac{n(Y_1 - \bar{Y})^2}{(n-1)\sum_i (Y_i - \bar{Y})^2}$ and $(Y_1, Y_2, \ldots, Y_n)$ are independent standard normal variables with mean zero and variance unity. It can be shown that the distribution of U^2 is beta with parameters $\frac{1}{2}$ and $(n-2)/2$ respectively and the p.d.f. of U is given by

$$g(u) = \frac{\Gamma\left(\dfrac{n-1}{2}\right)}{\Gamma(\frac{1}{2})\Gamma\left(\dfrac{n-2}{2}\right)}(1 - u^2)^{(n-4)/2}, \quad -1 < u < 1$$

and $= 0$ elsewhere

We can show by straight forward integration that the UMVUE of

$R(t \mid \mu, \sigma)$ is given by

$$\tilde{R}(t) = \tfrac{1}{2}\left[1 - I_{n(t-\bar{x})^2/(n-1s)^2}\left(\frac{1}{2}, \frac{n-2}{2}\right)\right] \quad \text{if } t - \bar{x} \geqslant 0$$

and

$$= \tfrac{1}{2}\left[1 + I_{n(t-\bar{x})^2/(u-1)s^2}\left(\frac{1}{2}, \frac{n-2}{2}\right)\right] \quad \text{if } t - \bar{x} < 0$$

where $I_w(a, b) = \dfrac{\Gamma(a+b)}{\Gamma(a)\Gamma(b)} \displaystyle\int_0^w x^{a-1}(1-x)^{b-1}\, dx$ is the well known incomplete beta integral. The above integral has been tabulated quite extensively. We refer the reader to such tables as Tables of Incomplete Beta integral originally edited and compiled by K. Pearson and reissued with a new introduction by E. S. Pearson and N. L. Johnson. For a given value of $n(t-\bar{x})^2/(n-1)s^2$, we can use package program also to evaluate the required incomplete beta integral.

Now suppose X is a two-parameter log-normal variate such that $\log X$ is normally distributed with mean μ and variance σ^2. Replacing $\log x$ by u, and $\bar{x}$ and s^2 by $\bar{u}$ and $s_u^2 = \displaystyle\sum_{i=1}^{n} (u_i - \bar{u})^2$ [as defined in (14), (15)] we may similarly obtain the MLE and UMVUE of $R(t)$ when one or both parameters are unknown.

EXAMPLE 3.4   (a) 15 items were put on test and the failure times (in hours) were:

$$13.4,\ 14.2,\ 28.8,\ 29.0,\ 29.8,\ 33.0,\ 37.8,\ 39.6,$$
$$43.4,\ 49.8,\ 54.8,\ 58.2,\ 67.4,\ 70.2,\ 91.2$$

Assuming the failure time distribution to be given by $N(\mu, \sigma = 20)$, compute $\tilde{R}_t$ at $t = 100$ and $t = 40$ hours.

(i) $t = 100$, $\bar{x} = 44.04$, $n = 15$

$$\frac{t - \bar{x}}{\sqrt{\dfrac{n-1}{n}}\,\sigma} = \frac{55.96}{19.32182} = 2.89621$$

$$\tilde{R}_t = 1 - \Phi(2.89621) = 0.00189$$

(ii) $t = 40$, $\dfrac{t - \bar{x}}{\sqrt{\dfrac{n-1}{n}}\,\sigma} = -0.2091$

$$\tilde{R}_t = -(-0.2091) = 0.58282$$

On the other hand, the corresponding MLE are given by $\tilde{R}(100) = 0.00258$ and $\tilde{R}_t(40) = 0.58004$.

(b) Following is the distribution of failure times of items under test:

$$4.4,\ 9.2,\ 44.8,\ 52.4,\ 62.8,\ 64.0,\ 65.2,\ 69.6,\ 110.4,\ 144.0$$

(in hours). Assuming the underlying distributions to be $N(\mu, \sigma^2)$, compute

$\tilde{R}_t$ at $t = 100$ and $t = 50$ hours.

(i) $t = 100$, $n = 10$, $s^2 = \dfrac{\sum\limits_{1}^{10} (x_i - \bar{x})^2}{10} = 1562.8176$

$$v^2 = \frac{(t - \bar{x})^2}{s^2(n - 1)} = 0.09902$$

$$I_{v^2}\left(\frac{1}{2}, \frac{n-2}{2}\right) = I_{v^2}\left(\frac{1}{2}, 4\right) = 0.62414$$

$$\tilde{R}_t = \tfrac{1}{2}(1 - 0.62414) = 0.1879$$

(ii) $t = 50$

$$v^2 = 0.01143$$
$$I_{v^2}(\tfrac{1}{2}, 4) = 0.23121$$
$$\tilde{R}_t = \tfrac{1}{2}(1 + 0.23121) = 0.6156$$

On the other hand, the corresponding MLE are given by $\hat{R}(100) = 0.17260$ and $\hat{R}(40) = 0.62580$.

EXAMPLE 3.5 Consider the sample from the log-normal distribution in Example 3.2. Given that $\sigma = 1$, estimate $R(t)$ at $t = 20$.

$$\hat{R}(t) = 1 - \Phi(\log t - \bar{u})$$
$$= 1 - \Phi(1.159)$$
$$= 0.123$$

$$\tilde{R}(t) = 1 - \Phi\left(\frac{\log t - \bar{u}}{\sqrt{\dfrac{n-1}{n}}}\right)$$

$$= 1 - \Phi(1.179)$$
$$= 0.119$$

Note that $\hat{R}(t)$ and $\tilde{R}(t)$ differ very slightly from the true value of $R(t \mid \mu, \sigma) = 0.1597$.

EXAMPLE 3.6 Given the data in Example 3.2, estimate the reliability function at time $t = 20$.

The MLE $\hat{R}(t) = 1 - \Phi\left\{\dfrac{\log t - \bar{u}}{\sqrt{s_u^2/n}}\right\}$

$$= 1 - \Phi\left\{\frac{2.9957 - 1.8363}{\sqrt{25.145/30}}\right\}$$

$$= 1 - \Phi(1.266)$$

$$= 0.103$$

$$v = \frac{\log t - \bar{u}}{\sqrt{(n-1)s_u^2/n}} = 0.2352$$

The UMVUE $R(t) = \frac{1}{2}\left[1 - \frac{1}{B(\frac{1}{2}, (n-2)/2)}\int_0^{v^2}\frac{(1-w)^{(n-4)/2}}{\sqrt{w}}\,dw\right]$

$\qquad\qquad\quad = \frac{1}{2}\left[1 - \frac{1}{B(\frac{1}{2}, 14)}\int_0^{0.5532} w^{1/2-1}(1-w)^{14-1}dw\right]$

$\qquad\qquad\quad = \frac{1}{2}(1 - 0.7891) = 0.105$

In this example also the $\hat{R}(t)$ and $\tilde{R}(t)$ do not differ very much. The true value of $R(t \mid \mu, \sigma) = 0.1597$.

Consider the bivariate normal distribution. The density function is given by

$$f(x, y \mid \mu_1, \mu_2, \sigma_1, \sigma_2, \rho) = \frac{1}{2\pi\sigma_1\sigma_2\sqrt{1-\rho^2}}\exp\left[-\frac{1}{2(1-\rho^2)}\left\{\left(\frac{x-\mu_1}{\sigma_1}\right)^2\right.\right.$$
$$\left.\left. + \left(\frac{y-\mu_2}{\sigma_2}\right)^2 - 2\rho\left(\frac{x-\mu_1}{\sigma_1}\right)\left(\frac{y-\mu_2}{\sigma_2}\right)\right\}\right] \qquad (21)$$

(i) Suppose σ_1, σ_2, ρ are known and without any loss of generality, let $\sigma_1 = \sigma_2 = 1$, $\rho = \rho_0$, $-1 < \rho_0 < 1$. It is well known that the sample mean vector $\overset{*}{\Theta} = (\bar{x}, \bar{y})$ is a complete sufficient statistic and

$$g_1(x, y \mid \overset{*}{\Theta}) = \frac{n}{2\pi(n-1)\sqrt{1-\rho_0^2}}\exp\left[-\frac{1}{2(1-\rho_0^2)}\left(\frac{n}{n-1}\right)\{(x-\bar{x})^2\right.$$
$$\left. + (y-\bar{y})^2 - 2\rho_0(x-\bar{x})(y-\bar{y})\}\right]$$

The UMVUE of $R(t)$ is given by

$$\tilde{R}(t) = 1 - \frac{1}{2\pi\sqrt{1-\rho_0^2}}\int_{-\infty}^{\sqrt{n/(n-1)}/(t-\bar{x})}\int_{-\infty}^{\sqrt{n/(n-1)}/(t-\bar{y})}\exp\left[-\frac{1}{2(1-\rho_0^2)}\{x^2+y^2\right.$$
$$\left. - 2\rho_0 xy\}\right]dxdy$$

When all the five parameters ($\mu_1, \mu_2, \sigma_1, \sigma_2, \rho$) are unknown, the complete sufficient statistic is given by $\overset{*}{\Theta} = (\bar{x}, \bar{y}, s_{11}, s_{12}, s_{22})$, where

$$s_{11} = \sum_{i=1}^n (x_i - \bar{x})^2, \; s_{12} = \sum_{i=1}^n (x_i - \bar{x})(y_1 - \bar{y}), \; s_{22} = \sum_{i=1}^n (y_i - \bar{y})^2$$

The UMVUE of $R(t)$ is now given by

$$\tilde{R}(t) = 1 - k\int_{\bar{x}-\sqrt{ns_{11}/(n-1)}}^t\int_{\bar{y}-\sqrt{ns_{22}/(n-1)}}^t g_2(x, y)\,dxdy$$

where

$$g_2(x, y \mid \overset{*}{\Theta}) = \begin{cases} (s_{11}s_{22} - s_{12}^2)^{-(n-3)/2}(s_{11}^*s_{22}^* - s_{12}^{*2}) & \text{if } s_{11}^*s_{22}^* - s_{12}^{*2} > 0, \\ 0 & \text{otherwise} \end{cases}$$

and $s_{11}^* = s_{11} - \dfrac{n}{n-1}(x - \bar{x})^2$, $s_{12}^* = s_{12} - \dfrac{n}{n-1}(x - \bar{x})(y - \bar{y})$,

$$s_{22}^* = s_{22} - \frac{n}{n-1}(y - \bar{y})^2, \quad k = \frac{n\Gamma(n/2)}{(n-1)\pi\Gamma(n-2)/2}$$

EXAMPLE 3.7 A random sample of 20 observations was generated from the bivariate normal distribution (2.1) with $\mu_1 = 5$, $\mu_2 = 8$, $\rho = 0.4$, $\sigma_1 = \sigma_2 = 1$. The true $R(t)$ as well as the UMVUE $\tilde{R}(t)$ were computed when: (i) μ_1, μ_2 are unknown, and (ii) all parameters are unknown.

SAMPLE

$$
x: \begin{cases}
6.4593, & 2.5958, & 4.3566, & 5.3056, & 4.5608, \\
4.2832, & 4.1212, & 4.5030, & 4.7463, & 5.3953, \\
6.3727, & 5.6172, & 4.9975, & 4.2544, & 3.5280, \\
5.4592, & 3.9433, & 6.0033, & 5.8480, & 3.6725.
\end{cases}
$$

$$
y: \begin{cases}
6.7700, & 6.1774, & 8.0511, & 6.9046, & 7.5105, \\
7.8583, & 7.1135, & 7.6804, & 9.5187, & 7.9812, \\
9.4351, & 7.1318, & 9.0031, & 7.2413, & 7.5259, \\
7.1041, & 8.4594, & 10.0538, & 8.0472, & 5.8871.
\end{cases}
$$

UMVUE of $R(t)$

t	$\tilde{R}(t \mid \mu_1, \mu_2$ unknown)	$\tilde{R}(t \mid$ all parameters unknown)	True $R(t)$
5	0.99790	0.99787	0.99875
6	0.96591	0.95603	0.97765
7	0.78634	0.76419	0.84174
8	0.40786	0.41671	0.50010
9	0.10399	0.12389	0.15866

(Samanta and Sinha, 1979)

Table 3.2 Values of z for Values of ψ and p

ψ	p									
	0.1	0.2	0.3	0.4	0.5	0.6	0.7	0.8	0.9	1.0
0.05	0.5449	0.6656	0.7811	0.9056	1.0498	1.2277	1.4646	1.8157	2.4459	4.35890
0.10	0.5374	0.6528	0.7916	0.8769	1.0076	1.1645	1.3652	1.6439	2.0854	3.00000
0.15	0.5294	0.6395	0.7417	0.8482	0.9665	1.1049	1.2758	1.5010	1.8267	2.38048
0.20	0.5210	0.6256	0.7272	0.8193	0.9262	1.0483	1.1944	1.3785	1.6271	2.00000
0.25	0.5119	0.6110	0.7002	0.7902	0.8865	0.9941	1.1193	1.2711	1.4653	1.73205
0.30	0.5022	0.5957	0.6786	0.7608	0.8473	0.9419	1.0492	1.1752	1.3292	1.52753
0.35	0.4918	0.5796	0.6562	0.7310	0.8083	0.8913	0.9832	1.0881	1.2116	1.36277
0.40	0.4806	0.5626	0.6329	0.7006	0.7694	0.8419	0.9205	1.0079	1.1076	1.22474
0.45	0.4685	0.5445	0.6087	0.6695	0.7303	0.7933	0.8603	0.9330	1.0138	1.05409
0.50	0.4552	0.5251	0.5833	0.6375	0.6909	0.7453	0.8020	0.8624	0.9278	1.00000
0.55	0.4405	0.5044	0.5565	0.6044	0.6508	0.6973	0.7450	0.7948	0.8476	0.94868
0.60	0.4244	0.4819	0.5281	0.5698	0.6097	0.6490	0.6887	0.7295	0.7718	0.81650
0.65	0.4063	0.4574	0.4976	0.5335	0.5672	0.6000	0.6326	0.6654	0.6991	0.73380
0.70	0.3858	0.4302	0.4647	0.4949	0.5228	0.5496	0.5758	0.6018	0.6280	0.65465
0.75	0.3621	0.3998	0.4285	0.4531	0.4757	0.4969	0.5174	0.5375	0.5574	0.57735
0.80	0.3342	0.3650	0.3878	0.4072	0.4246	0.4408	0.4562	0.4711	0.4857	0.50000
0.85	0.3001	0.3237	0.3408	0.3551	0.3677	0.3793	0.3901	0.4005	0.4104	0.42008
0.90	0.2561	0.2722	0.2836	0.2930	0.3011	0.3084	0.3152	0.3215	0.3276	0.33333
0.95	0.1921	0.2004	0.2062	0.2107	0.2147	0.2181	0.2212	0.2242	0.2269	0.22942

EXERCISES

3.1 Consider the distribution of the lives of 216 electric lamps out of 480 subjected to test.

Hours to failure (in units of 100 hours)	Frequency
1.05–2.05	8
2.05–3.05	10
3.05–4.05	14
4.05–5.05	24
5.05–6.05	45
6.05–7.05	55
7.05–8.05	60

Assume that the failure times are distributed as $N(\mu, \sigma^2)$. Using Gupta's (1952) methods for censored data compute the MLE of μ and σ and estimate the standard errors of these estimates. (The above is a grouped data and for calculation purposes use mid values of the interval to obtain individual observations.)

3.2 Ten television picture tubes were under test for 6000 hours. The first 8 failures occurred at 3220, 4300, 4870, 5132, 5472, 5634, 5876 and 5920 hours. Assuming the underlying life distribution to be $N(\mu, \sigma^2)$, obtain Plackett's (1959) estimators of μ and σ and estimate the standard errors of these estimates. Also obtain the MLE of μ and σ by using Gupta's results.

3.3 Suppose the distribution of the following failure times is given by $N(\mu, \sigma = 40)$. Compute $\widetilde{R}_t$ at $t = 100$ and $t = 50$ hours.

4.4	47.2	65.2
9.2	50.4	69.2
14.6	52.4	74.0
24.0	62.8	95.2
44.8	64.0	110.4

3.4 Let the distribution of the following failure times be given by $N(\mu, \sigma^2)$. Compute $\widetilde{R}_t$ at $t = 500$ and $r = 300$ hours.

100, 146, 176, 232, 244, 252, 364, 392, 476, 700.

3.5 Suppose the following failure times are distributed as

$$f(t \mid \mu, \sigma) = \frac{1}{\sqrt{2\pi}\sigma t} \exp\left\{-\frac{1}{2\sigma^2}(\log t - \mu)^2\right\}, \quad t > 0, \quad -\infty < \mu < \infty.$$

Compute the MLE and UMVUE of $R(t)$ at $t = 100$, given $\sigma = 2$.

4.6927	31.2182	78 3354	143.1653
14.2535	43.7722	93.3168	293.3024
15.9267	52.1957	118.5103	675.8694
18.2287	58.6155	119.5817	717.6629
20.1459	59.6802	138.5180	732.8932

3.6 Using the same data as in #3.5 and assuming μ and σ both unknown, compute $\widehat{R}(t)$ and $\widetilde{R}(t)$ at $t = 100$.

3.7 Let ξ_α denote the $100\alpha\%$ point of the standard normal distribution, i.e., $P(X \leqslant \xi_\alpha)$ $= \alpha = \int_{-\infty}^{\xi_\alpha} \phi(x)\,dx$. Then show that $100\alpha\%$ point of the $N(\mu, \sigma^2)$ is given by $\mu + \xi_\alpha \sigma$. Obtain the MLE of $\mu + \xi_\alpha \sigma$ and the UMVUE of $\mu + \xi_\alpha \sigma$. In the life testing context the $100\alpha\%$ point corresponds to the time before which $100\alpha\%$ of the items are likely to fail. (This problem is quite the reverse of the reliability estimation. In reliability estimation for a given time t_0 we want to estimate the probability that the item does not fail before t_0.)

3.8 A random sample of 50 was generated from $N(\mu = 5, \sigma = 2)$. A log-normal variate u is connected with a standard normal variate z by the relation $z = a \log (u = u_0) + b$. Obtain the moment estimates of a, b, u_0.

SAMPLE

7.1665060	5.4726666	6.8595063	3.4338160	1.0083030
2.7070744	1.0213795	5.7099263	2.3298625	5.4342824
4.2902986	2.0734258	2.3573799	8.0495314	6.6176362
6.1312164	9.6454727	3.9504556	6.7929762	4.2184889
2.3863370	7.2636821	2.6765919	2.5303836	−1.5212569
6.1066170	11.6905730	6.5044194	3.3230799	3.9720295
0.2277207	2.2482794	7.0012748	7.4090228	3.3436538
6.0455650	6.9503215	4.6385583	8.1741456	4.0647259
0.4108143	2.7314014	5.7277017	0 8264756	2.5592878
6.8207052	3.4685119	1.0407120	7.0235061	4.8138572

CHAPTER 4

Mixtures and Compound Distributions

In this chapter we will consider some models for failure time distributions which are more complex than those assumed thus far. In the previous chapters we assumed that the underlying population is a homogeneous one with the failure time distribution given by $F(x, \theta)$, where the form F is known but the parameter θ is unknown. The past experience as well as experimental constraints may suggest that the assumption of homogeneity may not hold and the underlying population may consist of several subpopulations, say $sp_1, sp_2, \ldots, sp_k$ mixed in proportions $p_1, p_2, \ldots, p_k$. Further, the failure time distribution in each subpopulation is given by $F_j(t)$, $j = 1, 2, \ldots, k$ with p.d.f.'s $f_j(t)$ respectively. A more common situation would be when $F_j(t)$, $j = 1, 2, \ldots, k$ all have same form F but they differ in parameters, i.e., $F_j(t) = F(t, \theta_j)$, $j = 1, 2, \ldots, k$. A random sample of size n is drawn from such a population giving $(X_1, X_2, \ldots, X_n)$ as failure times of the n items included in the sample. Two different types of situations commonly arise. In one case it is possible to assign (either after or at the start of the experiment) each unit to the appropriate subpopulation sp_j, $j = 1, 2, \ldots, k$, while in the other case such an information is not available. Therefore, in the first case the data would consist of the n failure times grouped according to the subpopulations,

$$\{(X_{11}, \ldots, X_{1n_1}), (X_{21}, \ldots, X_{2n_2}) \ldots (X_{k1}, \ldots, {}_{kn_k})\}$$

where it is assumed that $(n_1, n_2, \ldots, n_k)$ are the observed frequencies in the sample of the units belonging to subpopulations $sp_1, sp_2, \ldots, sp_k$ respectively. Assuming that $f(x, \theta_j)$ is the p.d.f. corresponding to the subpopulation sp_j, the likelihood of the sample is given by

$$L(x, n \mid \theta, p) = \frac{n!}{n_1! \, n_2! \ldots n_k!} \, p_1^{n_1} p_2^{n_2} \ldots p_k^{n_k} \prod_{j=1}^{k} \left\{ \prod_{i=1}^{n_j} f(x_{ji}, \theta_j) \right\} \quad (1)$$

In the other case where we can not assign a unit to a particular subpopulation sp_j, a straight forward computation shows that the failure time distribution of a unit is given by

$$G(t) = P\{X \leqslant t\} = \sum_{j=1}^{k} p_j F(t, \theta_j) \quad (2)$$

with the corresponding p.d.f. given by

$$g(x \mid \theta, p) = \sum_{j=1}^{k} p_j f(x, \theta_j) \tag{3}$$

In this case $(X_1, X_2, \ldots, X_n)$, the observed failure times of n units are regarded as a sample of size n from a population with failure time distribution given by $G(t \mid \theta, p)$ with the corresponding p.d.f. $g(x \mid \theta, p)$ defined in (2) and (3) above. The likelihood of the sample in this case would be given by

$$L(x \mid \theta, p) = \prod_{i=1}^{n} g(x_i \mid \theta, p) = \prod_{i=1}^{n} \left\{ \sum_{j=1}^{k} p_j f(x_i, \theta_j) \right\} \tag{4}$$

Both of these models will be referred to as mixtures and our initial objective is to estimate the parameters $(\theta_1, \theta_2, \ldots, \theta_k)$ when the mixing proportion $(p_1, p_2, \ldots, p_k)$ is assumed to be known or unknown, as the case may be.

Mixtures of Exponentials

The simplest case to study is $k = 2$, when there are only two subpopulations, sp_1 and sp_2 with mixing proportions p and $q = 1 - p$ and $f_1(u)$ and $f_2(u)$ are exponential densities with parameters θ_1 and θ_2 respectively. We will consider the case when items that fail can be classified and can be attributed to the appropriate subpopulations. Assuming that we have a complete sample the likelihood of the sample is given by

$$L(x \mid \theta_1, \theta_2, p) = \frac{n!}{n_1! \, n_2!} \frac{p^{n_1} q^{n_2}}{\theta_1^{n_1} \theta_2^{n_2}} \exp\left\{ -\sum_{i=1}^{n_1} x_{1i}/\theta_1 - \sum_{i=1}^{n_2} x_{2i}/\theta_2 \right\} \tag{5}$$

A straightforward derivation shows that the MLE of the parameters are

$$\hat{p} = \frac{n_1}{n}, \hat{q} = 1 - \hat{p}, \hat{\theta}_1 = \sum_{i=1}^{n_1} x_{1i}/n_1, \hat{\theta}_2 = \sum_{i=1}^{n_2} x_{2i}/n_2, n = n_1 + n_2$$

The above results can be generalized for any k, giving the following estimators

$$\hat{p}_j = n_j/n, \hat{\theta}_j = \sum_{i=1}^{n_j} x_{ji}/n_j, j = 1, 2, \ldots, k, n = \sum_{j=1}^{k} n_j$$

Mendenhall and Hader (1958) considered a more realistic and, therefore, a more difficult problem. The sources of complications are:

(i) an item can be attributed to the appropriate subpopulation only after it has failed, and

(ii) the experiment is time-censored and is terminated after a fixed time T.

We have n items on test and at the time of termination of the experiment the data consist of individual failure times, say r_1 items from sp_1 and of r_2 items from sp_2 and the information that $n - (r_1 + r_2)$ items survived beyond time T. To obtain the likelihood of the sample we note that the conditional p.d.f. of an item from sp_i $(i = 1, 2)$ given that it failed before time T is

$$f_i(t \mid T) = \frac{f_i(t)}{G_i(T)}, \quad 0 \leqslant t \leqslant T \tag{6}$$

where

$$f_i(t) = \frac{1}{\theta_i} \exp\left(-\frac{t}{\theta_i}\right)$$

and $\qquad G_i(t) = 1 - \exp\left(-t/\theta_i\right), \quad t > 0, \theta_i > 0, i = 1, 2$

Let $\qquad G(t) = pG_1(t) + qG_2(t) \tag{7}$

$\qquad\qquad$ = probability that a unit will fail before time t

and $\qquad R(t) = 1 - G(t)$

$$= pR_1(t) + qR_2(t) \tag{8}$$

$\qquad\qquad$ = probability that a unit will survive till time t

Following Mendenhall and Hader (1958), we will measure the time in units of T and change θ_i to $\beta_i = \theta_i/T$ $(i = 1, 2)$.

The probability that r_1 units from sp_1 fail before time T, r_2 units from sp_2 fail before time T and $(n - r)$ (where $r = r_1 + r_2$) units survive until time T is given by

$$\frac{n!}{r_1! \, r_2! \, (n - r)!} \, [pG_1(T)]^{r_1} [qG_2(t)]^{r_2} [R(T)]^{n-r} \tag{9}$$

Let x_{ij} denote the failure time of the jth item belonging to sp_i that failed before time T. Then in view of (6) and (9) the likelihood of the sample is given by

$$L(x_{11}, x_{12}, \ldots, x_{1r_1}; x_{21}, x_{22}, \ldots, x_{2r_2} \mid p, \beta_1, \beta_2)$$

$$= \frac{n!}{r_1! \, r_2! \, (n - r)!} \, p^{r_1} [G_1(T)]^{r_1} q^{r_2} \{G_2(T)\}^{r_2} [R(T)]^{n-r} \frac{\displaystyle\prod_{j=1}^{r_1} f_1(x_{ij})}{\{G_1(T)\}^{r_1}} \cdot \frac{\displaystyle\prod_{j=1}^{r} f_2(x_{2j})}{\{G_2(T)\}^{r_2}} \tag{10}$$

which simplifies to

$$L = (\text{Constant}) \frac{p^{r_1} q^{r_2}}{\beta_1^{r_1} \beta_2^{r_2}} \exp\left\{\frac{-r_1 \bar{x}_1}{\beta_1} - \frac{r_2 \bar{x}_2}{\beta_2}\right\} \left\{ p \exp\left(-\frac{1}{\beta_1}\right) \right.$$

$$\left. + q \exp\left(-\frac{1}{\beta_2}\right) \right\}^{(n-r)} \tag{11}$$

where

$$\bar{x}_1 = \frac{1}{r_1} \sum_{j=1}^{r_1} x_{1j}, \quad \bar{x}_2 = \frac{1}{r_2} \sum_{j=1}^{r_2} x_{2j}$$

Taking partial derivatives of log L with respect to β_1, β_2 and p we have

$$\frac{\partial}{\partial \beta_1} \log L = \frac{k(n-r)}{\beta_1^2} - \frac{r_1}{\beta_1} + \frac{r_1 \bar{x}_1}{\beta_1^2} \tag{12}$$

$$\frac{\partial}{\partial \beta_2} \log L = \frac{(1-k)(n-r)}{\beta_2^2} - \frac{r_2}{\beta_2} + \frac{r_2 \bar{x}_2}{\beta_2^2} \tag{13}$$

$$\frac{\partial}{\partial p} \log L = \frac{(n-r)k + r_1}{p} - \frac{(n-r)(1-k) + r_2}{1-p} \tag{14}$$

where

$$k = \frac{p \exp(-1/\beta_1)}{p \exp(-1/\beta_1) + q \exp(-1/\beta_2)}$$

The estimating equations are:

$$\hat{\beta}_1 = \bar{x}_1 + \hat{k}\, \frac{n-r}{r_1} \tag{15}$$

$$\hat{\beta}_2 = \bar{x}_2 + \frac{(1-\hat{k})(n-r)}{r_2} \tag{16}$$

$$\hat{p} = \frac{r_1 + \hat{k}(n-r)}{n} \tag{17}$$

$$\hat{k} = \frac{1}{1 + \dfrac{1-\hat{p}}{\hat{p}} \exp\left(\dfrac{1}{\hat{\beta}_1} - \dfrac{1}{\hat{\beta}_2}\right)} \tag{18}$$

Substituting $\hat{\beta}_1$, $\hat{\beta}_2$ and $\hat{p}$ in (18), we have an equation of the form

$$g(\hat{k}) = \hat{k}, \quad 0 \leqslant k \leqslant 1$$

Mendenhall and Hader devised a fairly complicated iterative scheme for solving the equation

$$g(\hat{k}) - \hat{k} = 0$$

Consider the truncated p.d.f.

$$f_i(x \mid x \leqslant 1) = \frac{1/\beta_i \exp(-x/\beta_i)}{1 - \exp(-1/\beta_i)}, \quad 0 \leqslant x \leqslant 1 \tag{19}$$

and $= 0$ otherwise

Let $\bar{x}_i$ denote the mean failure time based on r_i failures with p.d.f. given by (19). The MLE of β_i for the model (19) is a solution of

$$(\beta_i - \bar{x}_i)\left[\exp\left(\frac{1}{\beta_i}\right) - 1\right] = 1 \tag{20}$$

Plotting β_i as a function of $\bar{x}_i$ based on a sample from the p.d.f. (19), the MLE $\hat{\beta}_i$ can be obtained from Figure 1.

Pick the smaller of $(\bar{x}_1, \bar{x}_2)$, label this as sp_1 and obtain the corresponding $\hat{\beta}_{10}$ from Figure 1. Substituting $\hat{\beta}_{10}$ in (15), we have

$$\hat{k}_0 = \frac{(\hat{\beta}_{10} - \bar{x}_1)r_1}{n-r} \tag{21}$$

$$\widehat{\beta}_{20} = \bar{x}_2 + \frac{(1 - \widehat{k}_0)(n - r)}{r_2} \tag{22}$$

$$\widehat{p}_0 = \frac{r_1 + \widehat{k}_0(n - r)}{r} \tag{23}$$

$$g(\widehat{k}_0) = \frac{1}{1 + v_0} \tag{24}$$

where

$$v_0 = \frac{1 - \widehat{p}_0}{\widehat{p}_0} \exp \left(\frac{1}{\widehat{\beta}_{10}} - \frac{1}{\widehat{\beta}_{20}} \right) \tag{25}$$

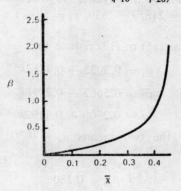

Fig. 1   Maximum likelihood estimate of β as a function of $\bar{x}$ based on a sample from a truncated exponential distribution. Measurements expressed in units of truncation time T (Reproduced by the kind permission of Mendenhall and Hader (1958) and the Editor of *Biometrika*).

If $g(\widehat{k}_0) - \widehat{k}_0 = 0$, $\widehat{k} = k_0$ and we are done. If $g(\widehat{k}_0) - \widehat{k}_0 = D_0 \neq 0$, consider the sign of D_0. Since $g(0) \geqslant 1$, $D_0 \lesseqgtr 0$ will imply $k \lesseqgtr k_0$

$$D = g(\widehat{k}) - \widehat{k}$$

$$= \frac{1}{1 + v} - \widehat{k}$$

$$\frac{dD}{d\widehat{k}} = - \left[\frac{1}{(1 + v)^2} \frac{dv}{d\widehat{k}} + 1 \right]$$

$$= - \left[1 + g^2(\widehat{k}) \frac{dv}{d\widehat{k}} \right]$$

and

$$d\widehat{k} = \frac{D_0}{1 + g^2(\widehat{k}) \dfrac{dv}{d\widehat{k}}}$$

where

$$D_0 = -dD$$

and

$$\frac{dv}{d\widehat{k}} = -v(n - r) \left[\left(\frac{1}{\widehat{q}} + \frac{1}{\widehat{p}} \right) \frac{1}{n} + \frac{1}{r_1 \widehat{\beta}_1^2} + \frac{1}{r_2 \widehat{\beta}_2^2} \right] \tag{26}$$

For the next iteration, write

$$\hat{k}_1 = \hat{k}_0 + d\hat{k}_0$$

$$= \hat{k}_0 + \frac{D_0}{1 + g^2(\hat{k}_0)\,\dfrac{d\nu}{d\hat{k}_0}}$$

Substituting $\hat{k}_1$ in the estimating equations (15) to (18) we obtain $\hat{\beta}_{11}$, $\hat{\beta}_{21}$, $\hat{p}_1$, $g(\hat{k}_1)$ and the process is repeated till the required accuracy is attained.

EXAMPLE 4.1   (Mendenhall and Hader, 1958) Consider a situation where $n = 369$, $r_1 = 107$, $r_2 = 218$, $T = 630$, $\bar{x}_1 = (\bar{t}_1/T) = 0.3034862$, $\bar{x}_2 = (\bar{t}_2/T) = 0.3644677$.

Estimating equations (15) to (17) reduce to

$$\hat{\beta}_1 = 0.3035 + 0.4112\hat{k} \tag{27}$$

$$\hat{\beta}_2 = 0.5663 - 0.2018\hat{k} \tag{28}$$

$$\hat{p} = 0.2900 + 0.1192\hat{k} \tag{29}$$

Pick the smaller of the two means $\bar{x}_1 = 0.303$ and obtain the first estimate

$$\hat{\beta}_{10} = 0.380$$

from Figure 1. Substituting in (21) to (24),

$$\hat{k}_0 = 0.186, \quad \hat{\beta}_{20} = 0.529, \quad \hat{p}_0 = 0.312, \quad g(\hat{k}_0) = 0.1779$$

$$D_0 = g(\hat{k}_0) - \hat{k}_0$$

$$= -0.0081 \text{ which implies that } 0 < \hat{k} < 0.186$$

From (25) and (26),

$$\nu_0 = \frac{0.688}{0.312}\exp\left(\frac{1}{0.380} - \frac{1}{0.529}\right) = 4.622$$

$$\frac{d\nu_0}{d\hat{k}_0} = -(4.622)(44)\left[\left(\frac{1}{0.688} + \frac{1}{0.312}\right)\frac{1}{369} + \frac{1}{(107)(0.380)^2} + \frac{1}{(218)(0.529)^2}\right]$$

$$= -19.04$$

$$d\hat{k}_0 = \frac{D}{1 + g^2(\hat{k}_0)\,\dfrac{d\nu_0}{d\hat{k}_0}} = \frac{-0.0081}{1 + (0.1779)^2(-19.04)} = -0.02$$

$$\hat{k}_1 = \hat{k}_0 + d\hat{k}_0 = 0.186 - 0.02 = 0.166$$

Using (27) to (29)

$$\hat{\beta}_{11} = 0.3718, \quad \hat{\beta}_{21} = 0.5358, \quad \hat{p}_1 = 0.3098$$

$$g(\widehat{k}_1) = \frac{1}{1 + \dfrac{1 - \widehat{p}_1}{\widehat{p}_1} \exp\left(\dfrac{1}{\widehat{\beta}_{11}} - \dfrac{1}{\widehat{\beta}_{21}}\right)} = 0.166$$

Thus, $D_1 = g(\widehat{k}_1) - \widehat{k}_1 = 0.000$ and hence $\widehat{\beta}_{11}$, $\widehat{\beta}_{21}$ and $\widehat{p}_1$ are the maximum likelihood estimators $\widehat{\beta}_1$, $\widehat{\beta}_2$, $\widehat{p}$ respectively

$$\widehat{\theta}_1 = T\widehat{\beta}_{11} = 234.2 \text{ hours}$$

$$\widehat{\theta}_2 = T\widehat{\beta}_{22} = 335.7 \text{ hours}$$

$$\widehat{p} = 0.3098$$

Record of Iteration

u	$\widehat{k}_u$	$\widehat{\beta}_{1u}$	$\widehat{\beta}_{2u}$	$\widehat{p}_u$	ν_u	$g(\widehat{k}_u)$	D_u
0	0.186	0.3800	0.5290	0.3120	4.622	0.1779	-0.0081
1	0.166	0.3718	0.5328	0.3098	5.024	0.1660	0.0000
2	0.167	0.3721	0.5326	0.3099	5.002	0.1666	-0.0004
3	0.165	0.3713	0.5330	0.3097	5.046	0.1654	0.0004

Reproduced by the kind permission of Mendenhall and Hader (1958) and the Editor of *Biometrika*.

From the above table it follows that the absolute value of the iterative error for $\widehat{k} < 0.001$.

Swamy and Doss (1961) obtained the MLE of the parameters (p, θ_1, θ_2) when items are put to test at different times. As pointed out by Bartholomew (1957), this is an important modification of great practical value in life testing experiments. Mendenhall and Hader as well as Swamy and Doss derived the (asymptotic) variance-covariance matrix of the estimators.

The problem of estimating (p, θ_1, θ_2) for the model (4) is extremely difficult and not much work has been done in this area. Rider (1961) obtained the moment estimators of the parameters for the complete sample situation and we consider his results in the following.

A simple calculation shows that the rth moment about the origin of the random variable X with the p.d.f.

$$f(x) = \frac{p}{\theta_1} \exp\left(-\frac{x}{\theta_1}\right) + \frac{1 - p}{\theta_2} \exp\left(-\frac{x}{\theta_2}\right) \tag{30}$$

is given by

$$\mu'_r = r! \, (p\theta_1^r + q\theta_2^r)$$

Hence,

$$\mu'_1 = p(\theta_1 - \theta_2) + \theta_2 \tag{31}$$

$$\frac{\mu'_2}{2} = p(\theta_1^2 - \theta_2^2) + \theta_2^2 \tag{32}$$

$$\frac{\mu'_3}{6} = p(\theta_1^3 - \theta_2^3) + \theta_2^3 \tag{33}$$

Eliminating p, we have

$$0 = \begin{vmatrix} \mu'_1 & \theta_1 - \theta_2 & \theta_2 \\ \mu'_2/2 & \theta_1^2 - \theta_2^2 & \theta_2^2 \\ \mu'_3/6 & \theta_1^3 - \theta_2^3 & \theta_2^3 \end{vmatrix}$$

$$= \begin{vmatrix} \mu'_1 & 1 & 1 \\ \mu'_2/2 & \theta_1 + \theta_2 & \theta_2 \\ \mu'_3/6 & \theta_1^2 + \theta_2^2 + \theta_1\theta_2 & \theta_2^2 \end{vmatrix}$$

$$= \begin{vmatrix} \mu'_1 & 0 & 1 \\ \mu'_2/2 & 1 & \theta_2 \\ \mu'_3/6 & \theta_1 + \theta_2 & \theta_2^2 \end{vmatrix}$$

$$= \mu'_3 - 3\mu'_2(\theta_1 + \theta_2) + 6\mu'_1\mu'_2\theta_2$$

Therefore,

$$\theta_1 = \frac{\mu'_3 - 3\mu'_2\theta_2}{3\mu'_2 - 6\mu'_1\theta_2} \tag{34}$$

Also, from (31) and (32),

$$\theta_1 = \frac{\mu'_2 - 2\mu'_1\theta_2}{2(\mu'_1 - \theta_2)} \tag{35}$$

(34) and (35) lead to a quadratic

$$6\theta^2(2\mu'^2_1 + \mu'_2) + 2\theta(\mu'_3 - 3\mu'_1\mu'_2) + 3\mu'^2_2 - 2\mu'_3\mu'_1 = 0 \tag{36}$$

Let m'_1, m'_2, m'_3 be the first three sample moments about the origin based on a random sample of n observations from (30). Substituting (m'_1, m'_2, m'_3) for (μ'_1, μ'_2, μ'_3) in (36) we obtain θ_1^* and θ_2^* as the two roots of the equation (36). Substituting θ_1^* and θ_2^* in (34) or (35) we have

$$p^* = \frac{m'_1 - \theta_2^*}{\theta_1^* - \theta_2^*}$$

It is possible that the roots of (36) will not be both positive or even real and the method may thus break down. However, in large samples this is not likely to occur. For details, see Rider (1961).

Mixtures of Weibull

Let the parent failure time distribution be made up of two subpopulations, each having the distribution function

$$1 - \exp\left(-\frac{t^{p_i}}{\theta_i}\right), \quad p_i, \theta_i, t > 0, 1 = 1, 2 \tag{37}$$

Let the subpopulations be mixed in the proportion $\alpha : \beta(= 1 - \alpha)$ and let us assume that an item can be attributed to the appropriate sub-population only after it has failed.

Suppose n items from the model (37) are subjected to a life testing experiment, the test terminating after a fixed time T and during this time r

units have failed, r_i from subpopulation i, $r = r_1 + r_2$. Let $(t_{i1}, t_{i2}, \ldots, t_{ir_i})$ be the ordered failure times and let $x_{ij} = t_{ij}/T \leqslant 1$. The likelihood of the sample is given by (10) where

$$G(t) = \alpha G_1(t) + \beta G_2(t)$$

and

$$R(t) = \alpha \exp\left(-\frac{t^{p_1}}{\theta_1}\right) + \beta \exp\left(-\frac{t^{p_2}}{\theta_2}\right)$$

The logarithm of the likelihood function is given by

$$
\begin{aligned}
\log L = \text{constant} &+ (n - r) \log\left[\alpha \exp\left(-\frac{T^{p_1}}{\theta_1}\right) + \beta \exp\left(-\frac{T^{p_2}}{\theta_2}\right)\right] \\
&+ r_1 \log \alpha + r_2 \log \beta + \sum_{j=1}^{r_1}\left[\log p_1 - \log \theta_1\right. \\
&+ (p_1 - 1) \log x_{1j} + p_1 \log T - \frac{(x_{1j}T)^{p_1}}{\theta_1}\Bigg] \\
&+ \sum_{j=1}^{r_2}\left[\log p_2 - \log \theta_2 + (p_2 - 1) \log x_{2j} + p_2 \log T - \frac{(x_{2j}T)^{p_2}}{\theta_2}\right]
\end{aligned}
\tag{38}
$$

Maximum Likelihood Estimators (MLE) of the Parameters

Consider the case when p_1, p_2, θ_1, θ_2, α are all unknown.

Differentiating (38) with respect to θ_1, θ_2, α, p_1 and p_2 in turn, we have

$$\frac{\partial \log L}{\partial \theta_1} = \frac{k(n - r)T^{p_1}}{\theta_1^2} - \frac{r_1}{\theta_1} + \frac{\sum_{j=1}^{r_1}(x_{1j}T)^{p_1}}{\theta_1^2}$$

$$\frac{\partial \log L}{\partial \theta_2} = \frac{(1 - k)(n - r)T^{p_2}}{\theta_2^2} - \frac{r_2}{\theta_2} + \frac{\sum_{j=1}^{r_2}(x_{2j}T)^{p_2}}{\theta_2^2}$$

$$\frac{\partial \log L}{\partial \alpha} = \frac{k(n - r) + r_1}{\alpha} - \frac{(1 - k)(n - r) + r_2}{\beta}$$

$$\frac{\partial \log L}{\partial p_1} = \frac{-k(n - r)T^{p_1} \log T}{\theta_1} + \frac{r_1}{p_1} + \sum_{j=1}^{r_1}\{\log (x_{1j}T)\}\left[1 - \frac{(x_{1j}T)^{p_1}}{\theta_1}\right]$$

$$\frac{\partial \log L}{\partial p_1} = \frac{-k(1-k)(n-r)T^{p_2} \log T}{\theta_2} + \frac{r_2}{p_2} + \sum_{j=1}^{r_2}\{\log (x_{2j}T)\}\left[1 - \frac{(x_{2j}T)^{p_2}}{\theta_2}\right]$$

where $k = \dfrac{1}{1 + \dfrac{\beta}{\alpha} \exp\left\{\dfrac{T^{p_1}}{\theta_1} - \dfrac{T^{p_2}}{\theta_2}\right\}}$

The estimating equations are

$$\hat{\alpha} = \frac{r_1}{n} + \hat{k}\left(\frac{n - r}{n}\right) \tag{39}$$

$$\hat{\theta_1} = \frac{\hat{k}(n - r)T^{\hat{p}_1} + \sum_{j=1}^{r_1}(x_{1j}T)^{\hat{p}_1}}{r_1} \tag{40}$$

$$\hat{\theta}_2 = \frac{(1 - \hat{k})(n - r)T^{\hat{p}_2} + \sum\limits_{j=1}^{r_2} (x_{2j}T)^{\hat{p}_2}}{r_2} \qquad (41)$$

where $\hat{p}_1$ and $\hat{p}_2$ are solutions of

$$g_1(\hat{p}_1) \equiv \frac{-(n - r)\hat{k}T^{\hat{p}_1}(\log T)}{\hat{\theta}_1} + \frac{r_1}{\hat{p}_1} + \sum_{j=1}^{r_1} \log (x_{1j}T)\left[1 - \frac{(x_{1j}T)^{\hat{p}_1}}{\hat{\theta}_1}\right] = 0 \qquad (42)$$

$$g_2(\hat{p}_2) \equiv \frac{-(n - r)(1 - \hat{k})T^{\hat{p}}(\log T)}{\hat{\theta}_2} + \frac{r_2}{\hat{p}_2} + \sum_{j=1}^{r_2} \log (x_{2j}T)\left[1 - \frac{(x_{2j}T)^{\hat{p}_2}}{\hat{\theta}_2}\right] = 0 \qquad (43)$$

and

$$\hat{k} = \frac{1}{1 + \dfrac{\hat{\beta}}{\hat{\alpha}} \exp\left\{\dfrac{T^{\hat{p}_1}}{\hat{\theta}_1} - \dfrac{T^{\hat{p}_2}}{\hat{\theta}_2}\right\}} \qquad (44)$$

Substituting arbitrary starting values $(\theta_1^*, \theta_2^*, \alpha^*, p_1^*, p_2^*)$ we solve the equations

$$g_1(\hat{p}_1) \equiv \frac{-(n - r)k^*T^{\hat{p}_1}(\log T)}{\theta_1^*} + \frac{r_1}{\hat{p}_1} + \sum_{j=1}^{r_1} \log T(x_{1j}T)\left[1 - \frac{(x_{1j}T)^{\hat{p}_1}}{\theta_1^*}\right] = 0 \qquad (45)$$

$$g_2(\hat{p}_2) \equiv \frac{-(n - r)(1 - k^*)T^{\hat{p}_2}(\log T)}{\theta_2^*} + \frac{r_2}{\hat{p}_2} + \sum_{j=1}^{r_2} \log (x_{2j}T)\left[1 - \frac{(x_{2j}T)^{\hat{p}_2}}{\theta_2^*}\right] = 0 \qquad (46)$$

for $\hat{p}_1$ and $\hat{p}_2$ by numerical iterations where

$$k^* = \frac{1}{1 + \dfrac{\beta^*}{\alpha^*} \exp\left\{\dfrac{T^{p_1^*}}{v_1^*} - \dfrac{T^{p_2^*}}{\theta_2^*}\right\}} \qquad (47)$$

Substitute k^* in (45) and iterate for $\hat{p}_1 = \hat{p}_{01}$ (say).

Substitute $[k^*]_{\hat{p}_{01}, p^*, \theta_1^*, \theta_2^*, \alpha^*}$ in (46) and iterate for $\hat{p}_2 = \hat{p}_{02}$.

Substitute $[k^*]_{\hat{p}_{01}, \hat{p}_{02}, \theta_1^*, \theta_2^*, \alpha^*}$ in (45) and iterate for $\hat{p}_1 = \hat{p}_{11}$.

Substitute $[k^*]_{\hat{p}_{11}, \hat{p}_{02}, \theta_1^*, \theta_2^*, \alpha^*}$ in (46) and solve for $\hat{p}_2 = \hat{p}_{12}$ and repeat the process till we obtain the solutions $(\hat{p}_1, \hat{p}_2) \equiv (p_{01}, p_{02})$ which satisfy (45) and (46). Once we obtain (p_{01}, p_{02}), the following iterative process (Deemer and Votaw, 1955; Mendenhall and Hader, 1958) may be used.

Consider the solutions $\hat{\theta}_1, \hat{\theta}_2$ in (40) and (41). Substituting in (44) we have an equation of the form

$$h(\hat{k}, \hat{p}_1, \hat{p}_2) - \hat{k} = 0 \qquad (48)$$

where $h(\ )$ is a non-negative bounded-function of $\hat{k}, \hat{p}_1, \hat{p}_2$.

Let

$$\widehat{\lambda}_i = \frac{\widehat{\theta}_i}{T^{\widehat{p}_i}}, \ \overline{y}_i = \frac{\sum_{j=1}^{r_i} x_{ij}^{\widehat{p}_i}}{r_i} \quad (i = 1, 2)$$

The estimating equations (40), (41) and (44) now become

$$\widehat{\lambda}_1 = \frac{(n-r)\widehat{k}}{r_1} + \overline{y}_1 \tag{49}$$

$$\widehat{\lambda}_2 = \frac{(n-r)(1-\widehat{k})}{r_2} + \overline{y}_2 \tag{50}$$

$$\widehat{k} = \frac{1}{1 + \dfrac{\widehat{\beta}}{\widehat{\alpha}} \exp\left(\dfrac{1}{\widehat{\lambda}_1} - \dfrac{1}{\widehat{\lambda}_2}\right)} \tag{51}$$

The MLE of λ_i are solutions of the equation

$$(\lambda_i - \overline{y}_i)\left\{\exp\left(\frac{1}{\lambda_i}\right) - 1\right\} = 1. \quad \text{[See (20)]} \tag{52}$$

Given $\widehat{p}_1 = p_{01}$, $\widehat{p}_2 = p_{02}$, pick $\overline{y} = \min(\overline{y}_1, \overline{y}_2)$ and without any loss of generality call it the subpopulation 1. Solve for the corresponding $\lambda = \lambda_0^*$ from (52), viz;

$$(\lambda_0^* - \overline{y})\left\{\exp\left(\frac{1}{\lambda_0^*}\right) - 1\right\} = 1$$

and $k = k_0$, the starting value of k, where

$$k_0 = \frac{r_1(\lambda_0^* - \overline{y})}{n - r}.$$

Estimate $\widehat{\alpha} = \alpha_0$ from (39), replacing k by k_0. Compare k_0 with $h(k_0, p_{01}, p_{02}) = \dfrac{1}{1 + v_0}$, $v_0 = \dfrac{1 - \alpha_0}{\alpha_0} \exp\left(\dfrac{1}{\lambda_{01}} - \dfrac{1}{\lambda_{02}}\right)$, where $(\lambda_{01}, \lambda_{02}) \equiv (\overline{\lambda}_1, \overline{\lambda}_2)$ with $\widehat{k} = k_0$ in (49) and (50). If $D_0 \equiv h(k_0, p_{01}, p_{02}) - k_0 = 0$, we are done and $(p_{01}, p_{02}, \alpha_0, \theta_{01} = \lambda_{01}T^{p_{01}}, \theta_{02} = \lambda_{02}T^{p_{02}})$ are the MLE of $(p_1, p_2, \alpha, \theta_1, \theta_2)$. If, however, $D_0 \neq 0$, compute $k_1 = k_0 + dk_0 = k_0 + \dfrac{D_0}{1 + h^2(k_0, p_{01}, p_{02})dv/dk_0}$ where $\dfrac{dv}{dk_0} = -(n-r)v_0\left[\dfrac{1}{n}\left(\dfrac{1}{\alpha_0} + \dfrac{1}{1-\alpha_0}\right) + \dfrac{1}{r_1\lambda_{01}^2} + \dfrac{1}{r_2\lambda_{02}^2}\right]$. (Refer to Mendenhall and Hader, 1958 and Sinha and Lee, 1979 for details.) Compare k_1 with $h(k_1, p_{01}, p_{02}) = \dfrac{1}{1 + v_1}$, $v_1 = \dfrac{1 - \alpha_1}{\alpha_1} \exp\left(\dfrac{1}{\lambda_{11}} - \dfrac{1}{\lambda_{12}}\right)$ and $(\alpha_1, \lambda_{11}, \lambda_{12}) \equiv (\widehat{\alpha}, \widehat{\lambda}_1, \widehat{\lambda}_2)$ with $\widehat{k} = k_1$ in (39), (49) and (50) respectively.

If $D_1 \equiv h(k_1, p_{01}, p_{02}) - k_1 = 0$, $(p_{01}, p_{02}, \alpha_1, \theta_{11}, \theta_{12})$ are the MLE of $(p_1, p_2, \alpha, \theta_1, \theta_2)$, where $(\theta_{11}, \theta_{12}) \equiv (\lambda_{11}T^{p_{01}}, \lambda_{12}T^{p_{02}})$.

If $D_1 \neq 0$, compute

$$k = k_1 + dk_1$$

$$= k_1 + \cfrac{D_1}{1 + h^2(k_1, p_{01}, p_{02})\cfrac{dv}{dk_1}}$$

and

$$\frac{dv}{dk_1} = -(n-r)v_1\left[\frac{1}{n}\left(\frac{1}{\alpha_1} + \frac{1}{1-\alpha_1}\right) + \frac{1}{r_1\lambda_{11}^2} + \frac{1}{r_2\lambda_{12}^2}\right]$$

and repeat the process. Suppose at some stage S_t we obtain a solution $k = k_t^*$ which satisfies the equation $h(k_t^*, p_{01}, p_{02}) - k_t^* = 0$ and let $(\theta_{t1}^*, \theta_{t2}^*, \alpha_t^*)$ be the corresponding MLE of $(\theta_1, \theta_2, \alpha)$. If $(\theta_{t1}^* - \theta_{(t-1),1}^*) = 0$ $(\theta_{t2}^* - \theta_{(t-1),2}^*) = 0$, $(\alpha_t^* - \alpha_{t-1}^*) = 0$, then $(\theta_{t1}^*, \theta_{t2}^*, \alpha_t^*, p_{01}, p_{02})$ are the MLE of $(\theta_1, \theta_2, \alpha, p_1, p_2)$. If not, substitute $(\theta_{t1}^*, \theta_{t2}^*, k_t^*)$ in (45) and (46) and solve for a new set $\hat{p}_1 = p_{t1}, \hat{p}_2 = p_{t2}$. Given p_{t1}, p_{t2}, pick the new $\bar{y} = \min(\bar{y}_1, \bar{y}_2)$, obtain the corresponding $\lambda = \lambda_1^*$ from (52) and $k_1^* = \cfrac{r_1(\lambda_1^* - \bar{y})}{n - r}$ and repeat the preceding process replacing (k_0, p_{01}, p_{02}) by (k_1^*, p_{t1}, p_{t2}). Suppose at some stage S_r we have the solution $k = k_r^*$ which satisfies the equation $h(k_r^*, p_{t1}, p_{t2}) - k_r^* = 0$. Let $(\theta_{r1}^*, \theta_{r2}^*, \alpha_r^*)$ be the corresponding MLE of $(\theta_1, \theta_2, \alpha)$. If $(\theta_{r1}^* - \theta_{(r-1),1}^*) = 0$, $(\theta_{r2}^* - \theta_{(r-1),2}^*)$ $= 0$, $(\alpha_r^* - \alpha_{(r-1)}^*) = 0$, then $(\theta_{r1}^*, \theta_{r2}^*, \alpha_r^*, p_{t1}, p_{t2})$ are the MLE of $(\theta_1, \theta_2, \alpha, p_1, p_2)$. If not, we continue the process till at some stage j we obtain the solutions $(\theta_{j1}^*, \theta_{j2}^*, \alpha_j^*, p_{*1}, p_{*2})$ which converge.

EXAMPLE 4.2 We generate 50 and 100 observations from the subpopulations 1 and 2 with $\theta_1 = 250$, $\theta_2 = 375$, $\alpha = \frac{1}{3}$, $p_1 = 1$, $p_2 = 1$.

Table 4.1 **Sample from Subpopulation 1, $n_1 = 50$**

211.4832	175.4812	50.1170	561.4687	617.2304
53.2834	99.7688	45.1315	83.2134	901.3647
127.8581	153.5822	393.0839	7.5080	247.0877
285.2824	20.2458	183.0842	207.1253	260.2941
547.3520	349.8488	370.9638	95.2040	27.6767
87.6436	910.9799	108.1031	251.9786	550.6755
44.9812	28.5336	121.5087	131.8994	460.7363
269.5200	40.8359	98.3575	294.4521	137.1630
548.7194	104.5854	5.1163	336.3872	1079.6620
70.8535	187.3612	353.6635	4.5376	13.2642

Table 4.2   Sample from Subpopulation 2, $n_2 = 100$

198.9965	198.2433	653.5061	558.2910	231.7297
12.6528	31.0559	410.5151	1096.6160	750.7640
385.8967	1337.1510	920.9221	1347.0810	53.2347
209.8570	163.8347	110.4131	119.6205	632.0517
192.7241	721.0405	55.1178	82.3419	72.5529
199.5567	446.7805	431.8437	62.2858	148.9624
168.4152	362.5832	141.8717	521.0239	239.3262
92.4667	110.2134	637.5363	381.9235	844.9895
431.2675	585.1977	347.8881	244.6254	75.5520
8.2283	124.9406	64.6036	144.1181	382.5886
258.4177	284.1938	918.0566	430.9328	14.4903
567.1623	117.1270	191.2685	94.0527	58.4981
218.8362	1991.4110	88.5051	228.2120	335.3662
62.1384	630.1501	27.5493	102.5474	71.1792
112.3546	58.7085	320.6489	514.8305	95.2729
290.7202	68.1725	257.5600	295.6079	587.4726
149.1163	704.4907	601.5539	513.1872	446.7385
71.8515	747.3177	181.0456	37.7467	17.7401
87.6151	320.8911	122.7632	111.9181	209.1385
1101.2350	70.1238	721.5310	219.8885	363.9187

Consider the data in Tables 4.1 and 4.2, truncated at $T = 650$ hours. We have $n = 150$, $r_1 = 47$, $r_2 = 86$. Using the arbitrary solutions $\theta_1^* = 258$, $\theta_2^* = 364$, $\alpha^* = 0.3025$, $p_1^* = 1$, $p_2^* = 1$ we solve (42) and (43) by numerical iterations described earlier and obtain $p_{01} = 1.010995$, $p_{02} = 1.014502$. Given (p_{01}, p_{02}), we obtain the solution $k = k_{01}^*$ and check if $D_{01}^* \equiv h(k_{01}^*, p_{01}, p_{02}) - k_{01}^* = 0$. We repeat the process described in (48)–(52). The first and the final iterations are given in Tables 4.3(a)–4.3(c).

Table 4.3 (a)   First Iteration with $\hat{p}_1 = 1.010995$, $\hat{p}_2 = 1.014502$

u	$\hat{k}_u$	$\hat{\theta}_{u1}$	$\hat{\theta}_{u2}$	$\hat{\alpha}_u$	ν_u	$h(\hat{k}_u, \hat{p}_1 \hat{p}_2)$	D_u
0	0.2260	270.4673	1780.0320	0.3390	17.2444	0.0548	−0.1712
1	0.0240	219.4484	389.9832	0.3160	8.3455	0.1070	0.0830
2	0.1786	285.4924	368.1550	0.8336	4.2751	0.1896	0.0110
3	0.2058	265.3657	364.3123	0.3367	3.8519	0.2061	0.0003
4	0.2065	265.5383	364.2158	0.3367	3.8420	0.2065	0.0000

Table 4.3 (b)   Final Iteration with $\hat{p}_1 = 0.981898$, $\hat{p}_2 = 1.031561$

u	$\hat{k}_u$	$\hat{\theta}_{u1}$	$\hat{\theta}_{u2}$	$\hat{\alpha}_u$	ν_u	$h(\hat{k}_u, \hat{p}_1\hat{p}_2)$	D_u
0	0.2610	235.5706	1907.5080	0.3429	14.6772	0.0638	−0.2354
1	0.0256	186.3589	431.7930	0.3162	7.5863	0.1165	0.1752
2	0.2008	222.9915	404.1768	0.3361	3.6700	0.2141	0.0354
3	0.2362	230.3985	398.5928	0.3401	3.2261	0.2366	0.0011
4	0.2373	230.6270	398.4204	0.3402	3.2135	0.2373	0.0000

Table 4.3 (c)   Maximum Likelihood Estimators of the Parameters $(p_1, p_2, \theta_1, \theta_2, \alpha)$

Parameters	Starting values	True values	MLE
p_1	1.0000	1.0000	0.9819
p_2	1.000	1.0000	1.0316
α	0.3025	0.3333	0.3402
θ_1	258.0000	250.0000	230.6272
θ_2	364.0000	375.0000	398.4204

The above computations were repeated with a new set of starting values $p_1^* = 1.2$, $p_2^* = 1.5$, $\theta_1^* = 240$, $\theta_2^* = 360$, $\alpha^* = 0.4$. Correct to four places, the resulting estimates are identical.

In general, the problem of estimation in the case of mixtures of distributions is quite complicated. For estimation of parameters from the mixture of normal distributions we refer to papers by Cohen (1967), Dick and Bowden (1973).

Compound Distributions

Now, suppose that an item may fail due to any one of the k mutually exclusive causes $(C_1, C_2, \ldots, C_k)$, i.e., if the item fails due to the cause C_i then it did not fail due to any other cause $C_j, j \neq i$. This situation can arise when an item under test has k different components and the item fails as soon as any one of the components fails. Thus k different components can be designated as k causes $(C_1, C_2, \ldots, C_k)$ and the first component to fail would be due to the cause C_i. We assume that the different components (causes) act independently with different failure time distributions $F_1(t), \ldots, F_k(t)$ and corresponding densities $f_1(t), \ldots, f_k(t)$. Denoting by $R(t)$, the reliability of the system, we have

$$R(t) = P\{X \geqslant t\} = \prod_{i=1}^{t} R_i(t) = \prod_{i=1}^{k} [1 - F_i(t)].$$

This follows from the fact that the system does not fail until time t if none of the k components fail. The probability of this event is $\prod_{i=1}^{k} [1 - F_i(t)]$, because of independence.

Let $H(t) = 1 - R(t)$ be the failure time distribution of the system. Then

$$H(t) = 1 - \prod_{i=1}^{k} [1 - F_i(t)]$$

Assuming that the failure time distribution of the ith component $F_i(t) = 1 - \exp(-t/\lambda_i)$, $i = 1, 2, \ldots, k$, we have

$$H(t) = 1 - \exp\left\{-t \sum_{i=1}^{k} \frac{1}{\lambda_i}\right\}$$

If we define $\frac{1}{\lambda} = \sum_{i=1}^{k} \frac{1}{\lambda_i}$, the failure time distribution of the system is given by

$$H(t) = 1 - \exp\{-t/\lambda\}, \text{ where } \frac{1}{\lambda} = \sum_{i=1}^{k} \frac{1}{\lambda_i} \tag{53}$$

and the corresponding density is given by

$$h(t) = \frac{1}{\lambda} \exp\{-t/\lambda\}, \ t > 0, \ \lambda > 0$$

This shows that a system consisting of k independent exponentially distributed components with means λ_i, has exponential failure time distribution with mean λ, where $\frac{1}{\lambda} = \sum_{i=1}^{k} \frac{1}{\lambda_i}$. Now in such types of problems it may be of interest to know $H_i(t)$ which is the probability of failure of the system due only to the ith component C_i. Clearly $H_i(t) = \text{Prob}\{Y_i < t$ and $Y_i < Y_j, j \neq i\}$, where Y_i denotes the failure time of the ith component of the system. In view of independence of components, we have

$$H_i(t) = \int \cdots_W \int f_1(x_1) \ldots f_k(x_k) \, dx_1 \ldots dx_k$$

where

$$W = \{0 < x_i < t, \text{ and } x_j > x_i, j \neq i\}$$

This immediately leads to

$$H_i(t) = \int_0^t \prod_{j \neq i} [1 - F_j(x_i)] f_i(x_i) \, dx_i \tag{54}$$

For the case of independent exponential components with means λ_i we have

$$\begin{aligned} H_i(t) &= \int_0^t \frac{1}{\lambda_i} \exp\left(-\frac{x_i}{\lambda_i}\right) \exp\left\{-x_i\left(\sum_{j \neq i} \frac{1}{\lambda_j}\right)\right\} dx_i \\ &= \int_0^t \frac{1}{\lambda_i} \exp\left\{-x_i\left(\sum_{j=1}^{k} \frac{1}{\lambda_j}\right)\right\} dx_i \\ &= \frac{\lambda}{\lambda_i}\left[1 - \exp\left(\frac{-t}{\lambda}\right)\right], \text{ where } \frac{1}{\lambda} = \sum_{j=1}^{k} \frac{1}{\lambda_i} \end{aligned} \tag{55}$$

Note that $H(t) = \sum_{i=1}^{k} H_i(t)$. From the above we can immediately calculate the conditional probability that the system failed due to the ith component, given that the system has failed. This probability $p_i = \lambda/\lambda_i$, $i = 1$,

$2, \ldots, k$. Note that $\sum_{i=1}^{k} p_i = \lambda \sum_{i=1}^{k} \frac{1}{\lambda_i} = 1$. We can also conclude that the conditional p.d.f. of the failure time of the system due to component is given by $h_i(t) = \frac{1}{\lambda_i} \exp(-t/\lambda)$, $i = 1, 2, \ldots, k$. Note also that $h(t) = \sum_{i=1}^{k} h_i(t)$.

We now use these results to obtain estimators of $(\lambda_1, \lambda_2, \ldots, \lambda_k)$ in the case of the following situation considered by Boardman and Kendell (1970a). For simplicity we assume $k = 2$. Suppose that a sample of n units is put on test and the test is to be terminated at a preassigned time T or earlier if all units fail before T. Assume that each unit on test consists of two independent exponential components with means λ_1, λ_2 respectively. Further, once the unit fails it is possible to determine whether the failure was due to component one or two. The data here would consist of r_1 units failed due to failure of component 1 with individual failure times (t_{11}, $t_{12}, \ldots, t_{1r_1}$), r_2 units failed due to failure of component 2 with individual failure times ($t_{21}, t_{22}, \ldots, t_{2r_2}$) and the fact $(n - r_1 - r_2)$ items did not fail before time T.

Let
$$r_1 \bar{t}_1 = \sum_{j=1}^{r_1} t_{1j}, \quad r_2 \bar{t}_2 = \sum_{j=1}^{r_2} t_{2j}, \quad r = r_1 + r_2$$

Then the likelihood of the sample is given by

$$L(\bar{t}_1, \bar{t}_2, r_1, r_2, | \lambda_1, \lambda_2) = \frac{n!}{r_1! \, r_2! \, (n - r)!} \frac{1}{\lambda_1^{r_1}} \exp\left\{-\frac{r_1 \bar{t}_1}{\lambda}\right\} \frac{1}{\lambda_2^{r_2}}$$

$$\exp\left\{-\frac{r_2 \bar{t}_2}{\lambda}\right\} \exp\left\{-\frac{(n - r)T}{\lambda}\right\} \qquad (56)$$

where $\frac{1}{\lambda} = \frac{1}{\lambda_1} + \frac{1}{\lambda_2}$. Writing $L(\lambda_1, \lambda_2)$ for the above likelihood we obtain

$$\log L(\lambda_1, \lambda_2) = \text{constant} - r_1 \log \lambda_1 - r_2 \log \lambda_2 - \frac{r_1 \bar{t}_1 + r_2 \bar{t}_2 + (n - r)T}{\lambda}$$

$$\frac{\partial}{\partial \lambda_1} \log L(\lambda_1, \lambda_2) = -\frac{r_1}{\lambda_1} + \frac{r_1 \bar{t}_1 + r_2 \bar{t}_2 + (n - r)T}{\lambda_1^2}$$

$$\frac{\partial}{\partial \lambda_2} \log L(\lambda_1, \lambda_2) = -\frac{r_2}{\lambda_2} + \frac{r_1 \bar{t}_1 + r_2 \bar{t}_2 + (n - r)T}{\lambda_2^2}$$

The MLE of λ_1, λ_2 are
$$\hat{\lambda}_i = \frac{r_1 \bar{t}_1 + r_2 \bar{t}_2 + (n - r)T}{r_i}$$

Boardman and Kendell have shown that $\hat{\lambda}_i$ is asymptotically unbiased for λ_i. Boardman (1973) discussed a more general case where the data are grouped into k classes, i.e., when the number of failures due to either cause is recorded for k specified time intervals.

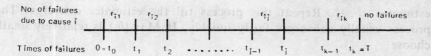

Probability of failure due to cause i between (t_{j-1}, t_j)

$$= \frac{1}{\lambda_i} \int_{t_{j-1}}^{t_j} \exp\left(-\frac{t}{\lambda}\right) dt$$

$$= \frac{\lambda}{\lambda_i} \left[\exp\left(-\frac{t_{j-1}}{\lambda}\right) - \exp\left(-\frac{t_j}{\lambda}\right) \right], \qquad j = 1, 2, \ldots, k; \; i = 1, 2$$

The likelihood may be written down as

$$L = \text{constant} \frac{\lambda^r}{\lambda_1^{r_1} \lambda_2^{r_2}} \prod_{j=1}^{k} \left\{ \exp\left(\frac{t_{j-1}}{\lambda}\right) - \exp\left(-\frac{t_j}{\lambda}\right) \right\}^{r_{1j}+r_{2j}} \exp\left\{ -\frac{(n-r)T}{\lambda} \right\}$$

where

$$r_i = \sum_{j=1}^{k} r_{ij} \text{ and } r = r_1 + r_2$$

$$\log L = \text{constant} + r_1 \log \frac{\lambda}{\lambda_1} + r_2 \log \frac{\lambda}{\lambda_2} + \sum_{j=1}^{k} (r_{1j} + r_{2j}) \log \left\{ \exp\left(-\frac{t_{j-1}}{\lambda}\right) \right.$$
$$\left. - \exp\left(-\frac{t_j}{\lambda}\right) \right\} - \frac{(n-r)T}{\lambda}.$$

Substituting $g_j = \dfrac{t_j + t_{j-1}}{2}$ and $h_j = (t_j - t_{j-1})$, Boardman obtained approximate MLE

$$\widehat{\lambda}_i^* = \left[\frac{\sum\limits_j (r_{1j} + r_{2j})g_j + (n-r)T - \dfrac{\sum\limits_j (r_{1j} + r_{2j})h_j^2}{12\widehat{\lambda}}}{r_i} \right] \tag{57}$$

Kendell (1963) suggested

$$\widehat{\lambda} = \frac{1}{2} \left[\widehat{\lambda}_0 + \left\{ \widehat{\lambda}_0^2 - \frac{\sum\limits_j (r_{1j} + r_{2j})h_j^2}{3r} \right\}^{1/2} \right] \tag{58}$$

where

$$\widehat{\lambda}_0 = \frac{1}{r} \left[\sum_j (r_{1j} + r_{2j})g_j + (n-r)T \right] \tag{59}$$

For the first iteration, substitute $\widehat{\lambda}$ from (58) to (57) and estimate λ_1^*, λ_2^*. For the second iteration, substitute $\widehat{\lambda} = \dfrac{\lambda_1^* \lambda_2^*}{\lambda_1^* + \lambda_2^*}$ in (57) and obtain new estimates λ_1^*, λ_2^*. Repeat this process till the estimates converge. The process usually converges fairly quickly. If Max $\{h_j\}$ is relatively small, choose

$$\hat{\lambda}_i = \frac{\sum_j (r_{1j} + r_{2j})g_j + (n - r)T}{r_i}$$

as initial estimators of λ, adjust these estimators by substituting $\lambda_0 = (r_1\hat{\lambda}_1 + r_2\hat{\lambda}_2)/r$ in (58) and solve for $\hat{\lambda}$. Substitute this $\hat{\lambda}$ in (57) and solve for λ_1, λ_2^*. Now, re-estimate $\hat{\lambda} = \dfrac{\hat{\lambda}_1^* \hat{\lambda}_2^*}{\hat{\lambda}_1^* + \hat{\lambda}_2^*}$ and substitute in (57) to obtain another set of new estimates $\hat{\lambda}_1^*, \hat{\lambda}_2$, and continue till the estimates converge. This procedure normally converges within four or five iterations.

EXAMPLE 4.3   Suppose two groups of 150 bulbs were placed on test. The data were grouped into 15 classes and the test was terminated after 1500 hours from the starting time. Assuming a compound exponential failure model, obtain approximate estimators of the parameter.

Group number	Hours to failure	Failure in type 1 (r_{1j})	Failure in type 2 (r_{2j})
1	0– 99	0	1
2	100– 199	1	1
3	200– 299	1	2
4	300– 399	2	2
5	400– 499	2	3
6	500– 599	3	3
7	600– 699	4	5
8	700– 799	10	35
9	800– 899	27	50
10	900– 999	44	32
11	1000–1099	34	8
12	1100–1199	11	2
13	1200–1299	3	1
14	1300–1399	2	1
15	1400–1499	1	1
		$r_1 = 145$	$r_2 = 147$

Group number	$r_{1j} + r_{2j}$	g_j	$(r_{1j} + r_{2j})g_j$
1	1	49.5	49.5
2	2	149.5	299.0
3	3	249.5	748.5
4	4	349.5	1398.0
5	5	449.5	2247.5
6	6	549.5	3297.0
7	9	649.5	5845.5
8	45	749.5	33727.5
9	77	849.5	65411.5
10	76	949.5	72162.0
11	42	1049.5	44079.0
12	13	1149.5	14943.5
13	4	1249.5	4998.0
14	3	1349.5	4048.5
15	2	1449.5	2899.0
Total	292		256154.0

$n = 300$, $T = 1500$, $r_1 = 145$, $r_2 = 147$, $r = 292$, $h_j = 100$, $j = 1, 2, \ldots, 15$. From (58) and (59), $\hat{\lambda}_0 = 918.3356$, $\hat{\lambda} = 917.4718$. Substituting in (57), $\hat{\lambda}_1^* = 1847.5088$, $\hat{\lambda}_2^* = 1822.3726$. Hence $\hat{\lambda} = \dfrac{\hat{\lambda}_1^* \hat{\lambda}_2^*}{\hat{\lambda}_1^* + \hat{\lambda}_2^*} = 917.4273$. Substituting in (57) $\hat{\lambda}_1^* = 1847.5087$, $\hat{\lambda}_2 = 1822.3725$ and $\hat{\lambda} = \dfrac{1847.5087 \times 1822.3735}{1847.5087 + 1822.3735} = 917.4275$. Substituting in (57), $\hat{\lambda}_1^* = 1847.5087$, $\hat{\lambda}_2^* = 1822.3725$ which converges.

Iteration	$\hat{\lambda}_1^*$	$\hat{\lambda}_2^*$
1	1847.5088	1822.3726
2	1847.5087	1822.3735
3	1847.5087	1822.3735

EXERCISES

4.1 A light bulb manufacturing company markets two brands of bulbs, *A* and *B*. A random sample of 54 bulbs were subject to test for 500 hours. 18 bulbs of brand *A* and 24 of brand *B* failed during the testing period and the failure times (in hours) were recorded as below:

A: 27, 30, 40, 50, 69, 73, 88, 94, 110, 133, 153, 163, 182, 201, 229, 238, 351, 495;

B: 38, 54, 59, 69, 71, 87, 92, 107, 114, 121, 136, 149, 152, 155, 176, 181, 193, 199, 235, 273, 315, 364, 416, 497.

Assuming that the life time t follows the mixed-exponential distribution and the likelihood function is given by (10) obtain the maximum likelihood estimates of p, θ_1, θ_2.

4.2 Suppose the following is a sample of 100 observations from a mixture model specified in (4.1). Compute the moment estimates of p, θ_1, θ_2.

SAMPLE

1	2	3	3	4	4	7	7	9	11
11	13	14	14	14	15	16	17	19	22
22	23	24	27	28	31	34	35	38	39
39	41	42	45	45	48	49	50	52	52
56	57	59	67	68	69	69	74	75	78
80	83	84	86	88	88	93	94	97	97
98	105	108	118	125	127	131	134	139	156
159	174	176	187	197	198	199	203	225	228
231	238	247	258	261	281	292	294	297	308
333	345	346	378	380	402	501	568	597	692

4.3 Treat the data in (4.2) as failure times of 100 units subject to a life-testing experiment having the same mixture model. Suppose the test was terminated after $T = 100$ hours. The failure times from the two subpopulations are:

Sample from subpopulation 1

2	3	4	4	7	9	11	13	14	14	15	17
19	22	24	28	31	35	38	39	41	45	48	52
57	59	68	69	74	75	80	84	88	93	97	98

Sample from subpopulation 2

1	3	7	4	14
16	22	23	27	34
39	42	45	49	50
52	56	67	69	78
83	86	88	94	97

Obtain the maximum likelihood estimates of p, θ_1, θ_2.

4.4 A company supplied 150 electronic units where each unit has two components A and B. The unit fails as soon as one of the components fails. Use the compound exponential model (64) to obtain the MLE of λ_1 and λ_2 when the number of failures and failure times were recorded as follows:

Hours to failure	No. of failures from company A	No. of failures from company B	Total
0–49	1	1	2
50–99	3	1	4
100–149	4	2	6
150–199	4	3	7
200–249	5	6	11
250–299	7	10	17
300–349	12	19	31
350–399	9	17	26
400–449	8	7	15
450–499	4	5	6
500–549	2	1	3
	59	72	131

4.5 (a) A unit U has two components and the unit fails when either of the components fail. Assume that the failure time distribution of the two types of components C_1 and C_2 have p.d.f.s

$$f_i(v) = \frac{1}{\lambda_i} \exp{(-v/\lambda_i)}, v, \lambda_i > 0, i = 1, 2$$

and the components are stochastically independent. Show that

(i) $P(U \leqslant t) = G(t) = 1 - \exp{\left\{-t\left(\frac{1}{\lambda_1} + \frac{1}{\lambda_2}\right)\right\}}$, is the d.f. of U, the failure time of the unit.

(ii) Show that the probability that the failure of the unit is due to C_2 and not to C_1 is given by $\lambda_1/(\lambda_1 + \lambda_2)$. (Hint: The event corresponds to the situation that $C_2 < C_1$.)

(b) Now suppose that the unit fails only when both components C_1 and C_2 fail Then show that

(i) $P(U \leqslant t) = G(t) = \left(1 - \exp{\left\{-\frac{t}{\lambda_1}\right\}}\right)\left(1 - \exp{\left\{-\frac{t}{\lambda_2}\right\}}\right)$, is the d.f. of U, the failure time of the unit.

(ii) Show that the probability that the failure of the unit is due to C_2 i.e., C_2 is the last unit to fail, is now given by $\lambda_2/(\lambda_1 + \lambda_2)$.

CHAPTER 5

Tests of Hypotheses and Confidence Intervals

In the previous chapters we considered the problem of estimation of parameters as well as the reliability function associated with various failure-time distributions commonly used in life testing experiments. In this chapter we will consider the problems of tests of hypotheses and the confidence intervals for the parameters in these models.

For example, in the single parameter exponential model with

$$f(x \mid \sigma) = \frac{1}{\sigma} \exp \left\{ -\frac{x}{\sigma} \right\}, \, x \geqslant 0, \, \sigma > 0$$

we may be interested in testing whether the mean life of an item, σ, is at least as large as σ_0 (a specification) against the alternative $\sigma < \sigma_0$. Another problem may be to test the reliability $R_{t_0}(\sigma) \geqslant p_0$ for specified t_0. Using the fact that $R_{t_0}(\sigma) = \exp \{ -(t_0/\sigma) \}$, this problem is equivalent to testing $\sigma \geqslant \sigma_0 = t_0/(-\log p_0)$ against the alternatives $\sigma < \sigma_0$. Problems of this nature occur quite frequently in quality control and in writing up warranties for the product.

For the basic theory and techniques of Tests of Hypotheses we refer to the Appendix 2 and to relevant material from the texts by Hogg and Craig (1978) and Mood, Graybill and Boes (1974). Generally H_0 denotes the null hypotheses and H_1 denotes the alternatives. We state here that the specification of H_0 and H_1 should be done quite carefully before the standard theory and techniques are applied to the problem at hand. As is well known, with a test procedure (or its equivalent critical region) are associated two types of error. The type I error is the probability of rejecting H_0 when H_0 is true and the type II error is the probability of accepting H_0 when H_0 is false or equivalently accepting H_0 when H_1 is true. It is customary to regard the type I error as more important than the type II error. One, therefore, tries to obtain a test procedure such that the type I error is bounded above by α (usually taken as 5% or 1%) and the type II error is as small as possible, or equivalently the power of the test (given by 1−type II error) is as large as possible. Thus H_0, the so-called null hypotheses, should be such that the rejection of H_0 when H_0 is true is more serious an error than accepting H_0 when H_0 is false. We illustrate these considerations by the following example.

Suppose a manufacturer of tyres wishes to test the hypotheses that the reliability of a tyre of a particular size, up to 2000 miles, is at least 90%. Assuming that the failure-time distribution follows the single parameter exponential distribution model, we have to test

$$R_{2000}(\sigma) \geqslant .90 \quad \text{or} \quad \sigma \geqslant \sigma_0 = \frac{2000}{(-\log .90)} = 18983 \simeq 19000 \text{ miles}$$

Note that here the distance measured in miles plays the role of time. Now the manufacturer may claim that $\sigma \geqslant \sigma_0$ when in fact $\sigma < \sigma_0$ or he may claim $\sigma < \sigma_0$ when in fact $\sigma \geqslant \sigma_0$. Assuming that the manufacturer has a long range interest in the market and that he is willing to stand behind the product, the error committed in claiming $\sigma \geqslant \sigma_0$, when in fact $\sigma < \sigma_0$ is much more serious than that committed in claiming $\sigma < \sigma_0$ when in fact $\sigma \geqslant \sigma_0$. Thus H_0 should be specified as $\sigma \leqslant \sigma_0$ and the alternative H_1 as $\sigma > \sigma_0$.

A closely connected problem with the tests of hypotheses is that of the construction of confidence intervals. For the basic theory and techniques we refer again to Hogg and Craig (1978) and Mood, Graybill and Boes (1974). We will illustrate the basic ideas by using the single parameter exponential model. Suppose we wish to construct one-sided confidence interval for the parameter σ, the expected life under the model. We then construct a function $T_L(x_1, \ldots, x_n)$ such that

$$P\{T_L(x_1, \ldots, x_n) \leqslant \sigma \mid \sigma\} = 1 - \alpha$$

holds for every value of $\sigma > 0$. The probability $(1 - \alpha)$ is called the confidence coefficient or level of the confidence interval which is usually specified in advance, generally at 90% or 95% or even 99%. For each observed value of $T_L(x_1, \ldots, x_n)$, we get the one-sided interval $(T_L(x_1, \ldots, x_n), \infty)$ such that the probability that this (random) interval covers the true value σ is exactly $(1 - \alpha)$. Similarly, we obtain a confidence interval of the type $(0, T_U(x_1, \ldots, x_n))$ of level $(1 - \alpha)$ by constructing a function $T_U(x_1, \ldots, x_n)$ such that

$$P\{\sigma \leqslant T_U(x_1, \ldots, x_n) \mid \sigma\} = 1 - \alpha$$

$T_L(x_1, \ldots, x_n)$ and $T_U(x_1, \ldots, x_n)$ are referred to as lower and upper confidence limits. For reliability and life testing problems one-sided intervals are more common but sometimes we need to construct two-sided intervals as well. These are obtained by constructing two functions $\{T'_L(x_1, \ldots, x_n), T'_U(x_1, \ldots, x_n)\}$ such that $T'_L(x_1, \ldots, x_n) \leqslant T'_U(x_1, \ldots, x_n)$, and $P\{T'_L(x_1, \ldots, x_n) \leqslant \sigma \leqslant T'_U(x_1, \ldots, x_n) \mid \sigma\} = 1 - \alpha$ holds for every $\sigma > 0$.

It may be mentioned here that sometimes in practice, for the given set of observations $(x_1, x_2, \ldots, x_n)$ the confidence interval statement such that

$$P\{T_L(x_1, \ldots, x_n) \leqslant \sigma \mid \sigma\} = 1 - \alpha$$

is interpreted as a statement that the probability that the unknown parameter σ lies in the interval $(T_L(x_1, \ldots, x_n), \infty)$ is $1 - \alpha$. Thus, we treat $T_L(x_1, \ldots, x_n)$ as fixed and σ as a random variable. However, as per the model, $T_L(x_1, \ldots, x_n)$ is a realized value of random variable and σ is constant. This is a controversial point and we will not go into the details here.

We also note the following connection between the confidence intervals and the tests of hypotheses. Suppose we have constructed $(T_L(x_1, \ldots, x_n), T_U(x_1, \ldots, x_n))$, a confidence interval of confidence coefficient $(1 - \alpha)$. Then,

$$P\{\sigma \notin (T_L(x_1, \ldots, x_n), T_U(x_1, \ldots, x_n)) \mid \sigma\} = \alpha$$

Suppose that we are testing $H_0 : \sigma = \sigma_0$ against $H_1 : \sigma \neq \sigma_0$ and the level of significance is specified to be α. Then we construct the two-sided confidence interval of confidence coefficient $(1 - \alpha)$ and if the observed confidence interval does not cover the hypothesized value σ_0, then we reject H_0. If we are testing $H_0 : \sigma \geqslant \sigma_0$ against $H_1 : \sigma < \sigma_0$ at level α we can use the one-sided confidence interval of level $1 - \alpha$ given by $(T_L(x_1, \ldots, x_n), \infty)$. Here

$$P\{T_L(x_1, \ldots, x_n) \leqslant \sigma \mid \sigma\} = 1 - \alpha$$

or

$$P\{\sigma < T_L(x_1, \ldots, x_n) \mid \sigma\} = \alpha$$

Therefore, if the boundary value $\sigma = \sigma_0$ lies below the observed value of $T_L(x_1, \ldots, x_n)$, or the confidence interval does not cover the value σ_0, then we reject H_0. Similar remarks are true for the problem of testing $H_0 : \sigma \leqslant \sigma_0$ against $H_1 : \sigma > \sigma_0$ at level α. Here we construct the $(1 - \alpha)$ level confidence interval $(0, T_U(x_1, \ldots, x_n))$ and reject H_0 if the boundary value σ_0 is larger than the observed value of $T_U(x_1, \ldots, x_n)$.

In view of the importance of the exponential model in the life testing and reliability problems, we will study this model first.

Exponential Distribution

We will consider the single parameter exponential distribution with p.d.f.

$$f(x, \sigma) = \frac{1}{\sigma} \exp\left(-\frac{x}{\sigma}\right), \; x > 0, \sigma > 0$$

and the failure censored samples, i.e., the case in which the test is terminated as soon as specified number of items, say r, fail. The results for the complete samples can be obtained by putting $r = n$.

Suppose we wish to test $H_0 : \sigma \geqslant \sigma_0$ against $H_1 : \sigma < \sigma_0$ at the level of significance α. Then we first consider the reduced problem of testing $H_0' : \sigma = \sigma_0$, the boundary value of σ under H_0, against the alternative $H_1' : \sigma = \sigma_1 < \sigma_0$ where σ_1 is any fixed value under the alternative H_1. The reduced problem can be solved by using the Neyman-Pearson lemma and the critical region is given by

$$\left(\frac{\sigma_1}{\sigma_0}\right)^r \exp\left\{\left(\frac{1}{\sigma_1} - \frac{1}{\sigma_0}\right) s_r\right\} \leqslant \text{constant} \tag{1}$$

where

$$s_r = \sum_{i=1}^{r} x_{(i)} + (n - r) x_{(r)} \text{ and } \{x_{(1)}, \ldots, x_{(r)}\}$$

are failure times of the r failed items arranged in the increasing order, or the r order statistics of the sample. Taking logarithms and dividing by $\left(\dfrac{1}{\sigma_1} - \dfrac{1}{\sigma_0}\right)$, which is positive, critical region given by (1) reduces to

$$S_r \leqslant c(\sigma_0, \sigma_1) \tag{2}$$

where $c(\sigma_0, \sigma_1)$ is a constant depending on σ_0. and σ_1. This constant is determined by the size condition, namely

$$P\{S_r \leqslant c(\sigma_0, \sigma_1) \mid \sigma_0\} = \alpha$$

where α is the specified level of significance. We have already seen that S_r/σ_0 is a gamma variable, or equivalently $2S_r/\sigma_0$ is a χ^2 variable with $2r$ degrees of freedom. Since extensive tables are available for $\chi^2(m)$ distributions, it is convenient to work with $Y = 2S_r/\sigma_0$. Let $\xi_p(k)$ denote the $100\,p\%$ point of the χ^2 distribution with k degrees of freedom, and $\xi_p(k)$ satisfy the relation

$$P\{\chi^2(k) \leqslant \xi_p(k)\} = p$$

A simple calculation then shows that if

$$P\left\{\frac{2S_r}{\sigma_0} < y \mid \sigma_0\right\} = \alpha$$

then $y = \xi_\alpha(2r)$. Thus, $c(\sigma_0, \sigma_1)$ in (2) is given by $\dfrac{\sigma_0}{2} \xi_\alpha(2r)$. Therefore, the critical region for the reduced problem is given by

$$S_r \leqslant \tfrac{1}{2} \sigma_0 \xi_\alpha(2r) \tag{3}$$

We note that the critical region depends only on σ_0 and is independent of σ_1. Thus (3) gives the critical region for the slightly larger problem, $H_0' : \sigma = \sigma_0$ against $H_1 : \sigma < \sigma_0$, since the critical region remains the same whatever be $\sigma_1 < \sigma_0$. If we show that the above critical region satisfies the size condition when $H_0' : \sigma = \sigma_0$ is enlarged to $H_0 : \sigma \geqslant \sigma_0$, then the critical region given by (3) will give us a size α test for the original problem $H_0 : \sigma \geqslant \sigma_0$ against $H_1 : \sigma < \sigma_0$. We now show that

$$P\{S_r \leqslant \tfrac{1}{2} \sigma_0 \xi_\alpha(2r) \mid \sigma\} \leqslant \alpha \text{ for } \sigma \geqslant \sigma_0 \tag{4}$$

Now $S_r \leqslant \tfrac{1}{2} \sigma_0 \xi_\alpha(2r)$ is equivalent to $\dfrac{2S_r}{\sigma} \leqslant \dfrac{\sigma_0}{\sigma} \xi_\alpha(2r)$. Therefore,

$$P\{S_r \leqslant \tfrac{1}{2}\sigma_0\xi_\alpha(2r) \mid \sigma\} = P\left\{Y \leqslant \frac{\sigma_0}{\sigma} \xi_\alpha(2r) \mid Y \sim \chi^2(2r)\right\}$$

$$= F_{2r}\left[\frac{\sigma_0}{\sigma} \xi_\alpha(2r)\right] \tag{5}$$

where $F_{2r}(a)$ is the distribution function of $\chi^2(2r)$ r.v. Now for $\sigma > \sigma_0$, $\frac{\sigma_0}{\sigma} \xi_\alpha(2r) < \xi_\alpha(2r)$ and, therefore,

$$F_{2r}\left[\frac{\sigma_0}{\sigma} \xi_\alpha(2r)\right] < F_{2r}[\xi_\alpha(2r)] = \alpha$$

This shows that the condition (4) is satisfied. Further, $F_{2r}\left[\frac{\sigma_0}{\sigma} \xi_\alpha(2r)\right]$ for $\sigma < \sigma_0$, gives the power of the test. It can be shown that this test is uniformly most powerful (UMP) test for the problem, i.e., for any other size α test the power of this test is maximum uniformly in $\sigma < \sigma_0$.

The same test can be obtained by using the method of likelihood ratio test where we reject H_0 if

$$\lambda(x) = \operatorname*{Sup}_{\sigma \geqslant \sigma_0} L(x \mid \sigma)/\operatorname*{Sup}_{\sigma > 0} L(x \mid \sigma) \leqslant \text{constant}$$

Note that here the MLE of σ under H_0, (i.e., $\sigma \geqslant \sigma_0$) is given by $\tilde{\sigma}_r = S_r/r$ if $S_r \geqslant r\sigma_0$ and $\tilde{\sigma}_r = \sigma_0$ otherwise. On the other hand the MLE of σ, when $\sigma > 0$, is $\hat{\sigma}_r = S_r/r$ without any restriction. Therefore,

$$\operatorname*{Sup}_{\sigma > 0} L(x, \sigma) = \binom{n}{r}\left(\frac{r}{S_r}\right)^r \exp(-r)$$

and

$$\operatorname*{Sup}_{\sigma \geqslant \sigma_0} L(x \mid \sigma) = \binom{n}{r}\left(\frac{r}{S_r}\right)^r \exp(-r) \qquad \text{if } S_r \geqslant r\sigma_0$$

$$= \binom{n}{r}\left(\frac{1}{\sigma_0}\right)^r \exp\left(-\frac{S_r}{\sigma_0}\right) \qquad \text{if } S_r < r\sigma_0$$

and the likelihood ratio statistics

$$\lambda(x) = 1 \qquad \text{if } S_r \geqslant r\sigma_0$$

$$= \frac{S_r}{r\sigma_0} \exp\left\{-\frac{S_r}{\sigma_0} + r\right\} \qquad \text{if } S_r < r\sigma_0$$

Our rejection rule is to reject H_0 if $\lambda(x) \leqslant$ constant where the constant is determined by the size condition, namely

$$P\{\lambda(x) \leqslant \mid H_0\} \leqslant \alpha$$

Thus, the critical region is a subset of

$$\lambda(x) = \left(\frac{S_r}{r\sigma_0}\right) \exp\left\{-\frac{S_r}{\sigma_0} + r\right\} \leqslant \text{a constant}$$

and we reject H_0 if $S_r < r\sigma_0$ and $S_r \leqslant$ a constant where the constant is chosen so that $P\{S_r \leqslant \text{constant} \mid H_0\} \leqslant \alpha$. Again we can show that this constant can be taken as $\frac{\sigma_0}{2} \xi_\alpha(2r)$.

In case we are testing $H_0 : \sigma \leqslant \sigma_0$ against $H_1 : \sigma > \sigma_0$, then we can show that the critical region now is given by $S_r \geqslant \text{constant} = \frac{\sigma_0}{2} \xi_{1-\alpha}(2r)$.

EXAMPLE 5.1 Consider the data of Example 1.2. Here $n = 60$ and $r = 10$ and the observed value of $S_{10} = 50,676$. Suppose we want to test the hypothesis about the reliability function $R_t(\sigma) = \exp(-t/\sigma)$ given by $H_0' : R_{600}(\sigma) \geqslant 0.90$ against $H_1' : R_{600}(\sigma) < 0.90$. Then this is equivalent to testing

$$H_0 : \sigma \geqslant \sigma_0 = \frac{600}{(-\log .90)} \simeq 5695 \text{ hours}$$

against

$$H_1 : \sigma < \sigma_0 = 5695 \text{ hours}$$

Here the critical region is given by $S_{10} \leqslant \frac{\sigma_0}{2} \xi_{.05}(20)$ if $\alpha = 0.05$. Thus, the critical value is given by $\frac{\sigma_0}{2} \xi_{.05}(20) = \frac{5695}{2} (10.85) \simeq 30,895$, where 10.85 is obtained from the tables of $\chi^2(20)$ distribution. Since we reject H_0 only if the observed value of $S_{10} \leqslant$ critical value, we do not reject H_0. Note that here the MLE of $R_{600}(\sigma) = .8887$ which is smaller than .90 but the difference is only .0113, which, as the test procedure shows, is not significant at 5% level of significance.

Suppose on the other hand we wish to test $H_0 : \sigma \leqslant 3000$ against $H_1 : \sigma > 3000$ at the same level $\alpha = 0.05$. Then the critical region is given by $S_{10} \geqslant \frac{3000}{2} \chi^2_{.95}(20) \simeq 47,175$. Here the observed value of S_{10} exceeds the critical value and we reject H_0.

Note that if in the above problems we change the level of significance from $\alpha = .05$ to $\alpha = .01$, then the critical values change to 26,302 and 56,349 respectively. Here we have a rather interesting situation in that we do not reject $\sigma \geqslant \sigma_0 \simeq 5695$ as well as we do not reject $\sigma \leqslant 3000$. Of course, logically this position appears untenable. We can interpret the situation as the case where we do not have strong enough evidence to reject both the hypotheses, even though the hypotheses are conflicting.

Next, we consider the problem of constructing one-sided or two-sided confidence intervals for σ. The confidence intervals are obtained by using a pivotal quantity, i.e., function of observations as well as the parameter for which we want to obtain a confidence interval. The defining property of such a pivotal quantity is that its distribution is known and remains the same whatever be the value of the parameter. In case of the single parameter exponential distribution, as we have seen earlier, $Y = 2S_r/\sigma$ is a pivotal quantity since the distribution of Y is $\chi^2(2r)$ under σ; thus

$$P\left\{\frac{2S_r}{\sigma} \leqslant a \mid \sigma\right\} = F_{2r}(a) \quad \text{for any } \sigma > 0 \tag{6}$$

The expression (6) immediately gives

$$P\left\{\frac{2S_r}{a} \leqslant \sigma \mid \sigma\right\} = F_{2r}(a) \quad \text{for any } \sigma > 0 \tag{7}$$

Thus, $(S_r/a, \infty)$ provides a one-sided confidence interval for σ, with the confidence coefficient given by $F_{2r}(a)$. Now if we want the confidence

level to be $1 - \alpha$, then we select a, such that $F_{2r}(a) = 1 - \alpha$ or $a = \xi_{1-\alpha}(2r)$.
Therefore, $T_L(x_1, \ldots, x_n) = \dfrac{2S_r}{\xi_{1-\alpha}(2r)}$ and $\left\{\dfrac{2S_r}{\xi_{1-\alpha}(2r)}, \infty\right\}$ is the required
one-sided confidence interval. If we want a confidence interval of the
type $\{0, T_U(x_1, \ldots, x_n)\}$, then we consider

$$P\left\{\frac{2S_r}{\sigma} \geqslant a \mid \sigma\right\} = 1 - F_{2r}(a)$$

or

$$P\left\{\sigma \leqslant \frac{2S_r}{a} \mid \sigma\right\} = 1 - F_{2r}(a)$$

We select a, such that $1 - F_{2r}(a) = 1 - \alpha$ or $a = \xi_\alpha(2r)$ and $T_U(x_1, \ldots, x_n)$
$= \dfrac{2S_r}{\xi_\alpha(2r)}$ gives us the upper confidence limit and $\left\{0, \dfrac{2S_r}{\xi_\alpha(2r)}\right\}$ provides us
with a one-sided confidence interval of level $1 - \alpha$.

EXAMPLE 5.2 Consider the data of Example 5.1. Suppose we want to
obtain one-sided confidence intervals of the type (T_L, ∞) or $(0, T_U)$ of level
$1 - \alpha = .95$. Noting that $S_{10} = 50,676$ and $\xi_{.95}(20) = 31.41$ and $\xi_{.05}(2)$
$= 10.85$, we obtain $T_L = \dfrac{2S_r}{\xi_{.95}(20)} = 3227$. Thus, $(3227, \infty)$ provides us a
95% confidence interval. We note that the value $\sigma_0 = 3000$ lies outside the
interval and we reject $\sigma \leqslant \sigma_0 = 3000$ at 5% level. Similarly $T_U = \dfrac{2S_r}{\xi_{.05}(20)}$
$\simeq 9341$ and $(0, 9341)$ gives us a 95% confidence interval. Again note that
for testing $H_0 : \sigma \geqslant \sigma_0$ against $\sigma < \sigma_0 \simeq 5695$, we observe that 5695 is
contained in the interval $(0, 9341)$ and we do not reject H_0.

Next, consider the two-sided confidence intervals for σ. Here we want
to obtain two functions $T_L(x_1, \ldots, x_n)$ and $T_U(x_1, \ldots, x_n)$ such that

$$P\{T_L \leqslant \sigma \leqslant T_U \mid \sigma\} = 1 - \alpha$$

holds for every $\sigma > 0$. Again we consider $Y = 2S_r/\sigma$ which is $\chi^2(2r)$ under
σ. Thus,

$$P\left\{a \leqslant \frac{2S_r}{\sigma} \leqslant b \mid \sigma\right\} = F_{2r}(b) - F_{2r}(a)$$

for any $\sigma > 0$. Clearly $(2S_r/b, 2S_r/a)$ provides us a confidence interval of
level $F_{2r}(b) - F_{2r}(a)$. If we now determine constants (a, b) such that
$F_{2r}(b) - F_{2r}(a) = 1 - \alpha$ for given α, we have confidence interval of level
$(1 - \alpha)$. Now there are many choices of (a, b) so that $F_{2r}(b) - F_{2r}(a)$
$= 1 - \alpha$. The choice of $a = \xi_{\alpha/2}(2r)$ and $b = \xi_{1-\alpha/2}(2r)$ corresponds to
equal tails criteria, i.e., we select (a, b) such that

$$P\{Y \leqslant a\} = P\{Y \geqslant b\} = \alpha/2$$

Another approach would be to consider the length of the confidence
interval, $2S_r\{(1/a) - (1/b)\}$. We then select (a, b) such that $F_{2r}(b) - F_{2r}(a)$
$= 1 - \alpha$ and $\{(1/a) - (1/b)\}$ is as small as possible. Using the standard

methods of minimization, such as Lagrangian multipliers, we can show that (a, b) are the solutions of

$$F_{2r}(b) - F_{2r}(a) = 1 - \alpha \quad \text{and} \quad a^2 f_{2r}(a) = b^2 f_{2r}(b) \tag{8}$$

where $f_{2r}(u)$ is the p.d.f. of a $\chi^2(2r)$ r.v. We have two equations in two unknowns and a unique solution (a, b) now can be obtained. Tate and Klett (1959) have tabulated solutions of (8) for $1 - \alpha = 0.90, 0.95, 0.99, 0.995, 0.999$ and the degrees of freedom $2r = 2, (1), 29$. For the values of $2r \geqslant 30$, we can use the normal approximation of $\chi^2(2r)$ and obtain $a = 2r + Z_{\alpha/2}\sqrt{4r}$ and $b = 2r + Z_{1-\alpha/2}\sqrt{4r}$, where Z_p denotes the $100\,p\%$ point of the standard normal distribution.

The problem for obtaining the confidence intervals for the reliability function $R_{t_0}(\sigma)$ can be solved by noting that $R_{t_0}(\sigma)$ is an increasing function of σ. Thus $\psi_1(x_1, \ldots, x_n) \leqslant \exp\left(-\dfrac{t_0}{\sigma}\right)$ is equivalent to $\sigma \geqslant \dfrac{t_0}{-\log \psi_1(x_1, \ldots, x_n)}$ and $\psi_2(x_1, \ldots, x_n) \geqslant \exp\left(-\dfrac{t_0}{\sigma}\right)$ is equivalent to $\sigma \leqslant \dfrac{t_0}{-\log \psi_2(x_1, \ldots, x_n)}$.

Therefore, the expression

$$P\left\{\psi_1(x_1, \ldots, x_n) \leqslant \exp\left(-\frac{t_0}{\sigma}\right) \leqslant \psi_2(x_1, \ldots, x_n)\,\Big|\,\sigma\right\} = 1 - \alpha \tag{9}$$

is equivalent to

$$P\left\{\frac{t_0}{-\log \psi_1} \leqslant \sigma \leqslant \frac{t_0}{-\log \psi_2}\,\Big|\,\sigma\right\} = 1 - \alpha$$

It immediately follows that $2S_r/b = t_0/-\log \psi_1$ and $2S_r/a = t_0/-\log \psi_2$ where (a, b) are given by (8) and we have

$$\psi_1(x_1, \ldots, x_n) = \exp\left(-\frac{bt_0}{2S_r}\right) \quad \text{and} \quad \psi_2(x_1, \ldots, x_n) = \exp\left(-\frac{at_0}{2S_r}\right)$$

Therefore, $\{\exp(-bt_0/2S_r), \exp(-at_0/2S_r)\}$ provides us with the confidence intervals for $\exp(-t_0/\sigma)$.

EXAMPLE 5.3 Again consider the data for Example 5.1 and we want to obtain two-sided confidence interval for σ at level 95%. Here $2r = 20$ and therefore, using Tate and Klett (1959) tables, we find $a = 10.6119$ and $b = 39.5611$. The observed $S_{10} = 50,676$ and the 95% confidence interval is given by $(2561.91, 9550.79)$ which for all practical purposes can be taken to be $(2562, 9551)$.

Similarly, if we want to obtain a 95% confidence interval for $R_t(\sigma)$ at $t = 600$ hours, we compute $\exp(-at_0/2S_r) = .9391$ and $\exp(-bt_0/2S_r) = .7912$ and $(.7912, .9391)$ gives us the 95% confidence interval for $R_{600}(\sigma)$.

Consider the two-parameter exponential distribution given by

$$f(x \mid \mu, \sigma) = \frac{1}{\sigma} \exp\left\{-\frac{(x - \mu)}{\sigma}\right\}, \quad x \geqslant \mu, \ -\infty < \mu < \infty, \ \sigma > 0$$

Suppose we want to test the hypotheses about σ such as $H_0 : \sigma \geqslant \sigma_0$ against $H_1 : \sigma < \sigma_0$. Note that μ is now unknown but neither H_0 nor H_1 specifies μ or makes any statement about μ. In such a situation the parameter μ is usually referred to as a nuisance parameter. The problem of developing tests of hypotheses when a nuisance parameter is involved is generally quite difficult. One should as a first attempt try to obtain a likelihood ratio test and check whether such a test has desirable properties. In this problem the MLE's of (μ, σ) are given by $x_{(1)}$ and S_r/r, where

$$S_r = \sum_{i=1}^{r} \{(x_{(i)} - x_{(1)}\} + (n - r)(x_{(r)} - x_{(1)})$$

while the MLE's of (μ, σ) under H_0 are given by $x_{(1)}$ and $\hat{\sigma}_0 = S_r/r$ if $S_r/r \geqslant \sigma_0$ and $\hat{\sigma}_0 = \sigma_0$ otherwise. Proceeding in the same way, as was done in the one-parameter case, we can deduce that the likelihood ratio test gives the critical region $S_r \leqslant$ constant. Noting that $2S_r/\sigma$ is $\chi^2(2r - 2)$ for any (μ, σ), we can determine this constant subject to the size condition $P\{S_r \leqslant \text{constant} \mid H_0\} \leqslant \alpha$. The constant is here given by $(\sigma_0/2)\xi_\alpha(2r - 2)$. We note here that the nuisance parameter μ did not cause any problem perhaps due to the fact that S_r/r is a good estimate of σ and moreover the distribution of S_r/r depends only on σ and is independent of μ.

Suppose we want to test $H_0 : \mu \leqslant \mu_0$ against $H_1 : \mu > \mu_0$ where σ is unspecified. Let Ω_{H_0} and Ω_{H_1} denote the parameter space under H_0 and H_1 respectively and let $\Omega = \Omega_{H_0} \cup \Omega_{H_1}$. Now as seen in Chapter 1, the likelihood of the sample is given by

$$L(x \mid \mu, \sigma) = \text{constant} \frac{1}{\sigma^r} \exp \left\{ - \sum_{i=1}^{r} \frac{(x_{(i)} - \mu)}{\sigma} - \frac{(n - r)(x_{(r)} - \mu)}{\sigma} \right\}$$

$$= \text{constant} \frac{1}{\sigma^r} \exp \left\{ - \frac{S_r + n(x_{(1)} - \mu)}{\sigma} \right\} \qquad (11)$$

Now the MLE's of μ and σ, when $(\mu, \sigma) \in \Omega$, are given by $\hat{\mu} = x_{(1)}$, $\hat{\sigma} = S_r/r$ and, therefore,

$$\sup_{\Omega} L(x \mid \mu, \sigma) = \text{constant} \frac{1}{\hat{\sigma}^r} \exp (-r) \qquad (12)$$

Now consider the maximization of $L(x \mid \mu, \sigma)$ when $(\mu, \sigma) \in \Omega_{H_0}$. Following the same methods as of Chapter 1, we first fix σ and maximize (11) subject to $\mu \leqslant \mu_0$ and also subject to $x_{(1)} \geqslant \mu$. Again in order to maximize (11) for fixed S_r, $x_{(1)}$ and σ, we must select the maximum possible value of μ subject to $\mu \leqslant \mu_0$ and $\mu \leqslant x_{(1)}$. Two cases arise if $x_{(1)} > \mu_0$, then $\tilde{\mu} = \mu_0$ and if $x_{(1)} \leqslant \mu_0$, then $\tilde{\mu} = x_{(1)}$. Again maximizing $L(x \mid \mu, \sigma)$ for $\sigma > 0$, we obtain

$$\tilde{\sigma} = \frac{S_r + n(x_{(1)} - \tilde{\mu})}{r} \text{ or } \tilde{\sigma} = \frac{S_r}{r} \quad \text{if } x_{(1)} \leqslant \mu_0$$

and

$$\tilde{\sigma} = \frac{S_r + n(x_{(1)} - \mu_0)}{r} \qquad \text{if } x_{(1)} > \mu_0$$

Therefore, we have

$$\sup_{\Omega_{H_0}} L(x \mid \mu, \sigma) = \text{constant } \frac{1}{\sigma^r} \exp(-r) \tag{13}$$

A straightforward computation shows that the likelihood ratio test statistics

$$\sup_{\Omega_{H_0}} L(x \mid \mu, \sigma) / \sup_{\Omega} L(x \mid \mu, \sigma) = \lambda(x)$$

where
$$\lambda(x) = 1 \qquad\qquad \text{if } x_{(1)} \leqslant \mu_0$$

$$= \left[\frac{S_r}{S_r + n(x_{(1)} - \mu_0)} \right] \quad \text{if } x_{(1)} > \mu_0$$

The rejection region is defined by $\lambda(x) \leqslant c_1$, where c_1 is obtained so that $P\{\lambda(x) \leqslant c_1 \mid H_0\} \leqslant \alpha$, where α is the level of significance. Now $\lambda(x) \leqslant c_1$ is equivalent to $\dfrac{S_r}{S_r + n(x_{(1)} - \mu_0)} \leqslant c_2$. Note that since $0 < \alpha < 1$, if $\lambda(x) = 1$, we accept H_0. Again $\dfrac{S_r}{S_r + n(x_{(1)} - \mu_0)} \leqslant c_2$ is equivalent to $\dfrac{n(x_{(1)} - \mu_0)}{S_r} \geqslant c_3$. Thus we reject H_0 if

$$T = \frac{n(x_{(1)} - \mu_0)}{S_r} \geqslant c_3 \tag{14}$$

where c_3 is such that

$$P\left\{ \frac{n(x_{(1)} - \mu_0)}{S_r} \geqslant c_3 \mid H_0 \right\} \leqslant \alpha$$

Consider the r.v.

$$T' = \frac{n(x_{(1)} - \mu_0)}{S_r} = \frac{2n(x_{(1)} - \mu_0)}{\sigma} \Big/ \frac{2S_r}{\sigma} = U/V$$

Now when $\mu = \mu_0$, we know that U is $\chi^2(2)$ and V is $\chi^2(2r - 2)$ and U and V are independent. Thus, we can use the ratio

$$T^0 = \frac{2(r - 1)}{2} U \Big/ V = \frac{n(x_{(1)} - \mu_0)}{S_r/(r - 1)}$$

which is distributed as F with 2, and $2(r - 1)$ degrees of freedom. Thus when $\mu = \mu_0$ we determine $c_3(\mu_0, \alpha)$ so that

$$P\{T^0 \geqslant c_3(\mu_0, \alpha) \mid \mu_0, \sigma\} = \alpha$$

Therefore, $c_3(\mu_0, \alpha) = F_{1-\alpha}(2, 2r - 2)$ where $F_p(v_1, v_2)$ is the $100p\%$ point of the standard F r.v. with v_1 and v_2 degrees of freedom. Hence, we reject H_0 if

$$x_{(1)} \geqslant \mu_0 + \frac{S_r}{n(r - 1)} F_{1-\alpha}(2, 2r - 2) \tag{15}$$

To claim that (15) gives the size α test, we must show that

$$P\left\{ x_{(1)} \geqslant \mu_0 + \frac{S_r}{n(r - 1)} F_{1-\alpha}(2, 2r - 2) \mid \mu \leqslant \mu_0, \sigma \right\} \leqslant \alpha$$

This can be proved by noting that $\dfrac{n(x_{(1)} - \mu)}{S_r/(r - 1)}$ is $F(2, 2 - r)$ under (μ, σ). Therefore,

$$P\left\{x_{(1)} \geqslant \mu + \frac{S_r}{n(r - 1)} F_{1-\alpha}(2, 2r - 2) \mid \mu, \sigma\right\} = \alpha \qquad (16)$$

for all $(\mu, \sigma) \in \Omega$. Since $\mu_0 > \mu$, we have

$$\mu_0 + \frac{S_r}{n(r - 1)} F_{1-\alpha}(2, 2r - 2) > \mu + \frac{S_r}{n(r - 1)} F_{1-\alpha}(2, 2r - 2)$$

for any S_r and therefore,

$$P\left\{x_{(1)} \geqslant \mu_0 + \frac{S_r}{n(r - 1)} F_{1-\alpha}(2, 2r - 2) \mid \mu, \sigma\right\}$$

$$\leqslant P\left\{x_{(1)} \geqslant \mu + \frac{S_r}{n(r - 1)} F_{1-\alpha}(2, 2r - 2) \mid \mu, \sigma\right\} = \alpha$$

The fact that $\dfrac{n(x_{(1)} - \mu)}{S_r/(r - 1)}$ has $F(2, 2r - 2)$ distribution for any (μ, σ), can be used to obtain confidence intervals for μ. Thus (16) can be rewritten as

$$P\left\{x_{(1)} < \mu + \frac{S_r}{n(r - 1)} F_{1-\alpha}(2, 2r - 2) \mid \mu, \sigma\right\} = 1 - \alpha$$

or

$$P\left\{x_{(1)} - \frac{S_r}{n(r - 1)} F_{1-\alpha}(2, 2r - 2) < \mu \mid \mu, \sigma\right\} = 1 - \alpha \qquad (17)$$

which shows that $\left(x_{(1)} - \dfrac{S_r}{n(r - 1)} F_{1-\alpha}(2, 2r - 2), \infty\right)$ gives us a one-sided confidence interval of confidence coefficient $(1 - \alpha)$. Similarly using

$$P\left\{\frac{n(x_{(1)} - \mu)}{S_r/(r - 1)} > F_\alpha(2, 2r - 2) \mid \mu, \sigma\right\} = 1 - \alpha$$

for all $(\mu, \sigma) \in \Omega$ we obtain

$$P\left\{\mu < x_{(1)} - \frac{S_r}{n(r - 1)} F_\alpha(2, 2r - 2) \mid \mu, \sigma\right\} = 1 - \alpha$$

$$\text{for all } (\mu, \sigma) \in \Omega \qquad (18)$$

Hence, $\left(-\infty, x_{(1)} - \dfrac{S_r}{n(r - 1)} F_\alpha(2, 2r - 2)\right)$ provides the $100(1 - \alpha)\%$ confidence interval of the type $(-\infty, T_U)$. We can use (18) to obtain a test for $H_0 : \mu \geqslant \mu_0$ against $H_1 : \mu < \mu_0$. Here our rejection rule is to reject H_0 if

$$x_{(1)} < \mu_0 + \frac{S_r}{n(r - 1)} F_\alpha(2, 2r - 2)$$

Two-sided confidence intervals for μ can also be obtained in a similar manner. Thus, we have for any constants (a, b), $a < b$

$$P\left\{a \leqslant \frac{n(x_{(1)} - \mu)}{S_r/(r - 1)} \leqslant b\right\} = H_{2 \cdot 2r-2}(b) - H_{2 \cdot 2r-2}(a)$$

where $H_{2 \cdot 2r-2}$ is the distribution function of F r.v. with 2 and $(2r-2)$ degrees of freedom. If we select (a, b) such that $H_{2 \cdot 2r-2}(b) - H_{2 \cdot 2r-2}(a) = 1 - \alpha$, we will have a confidence interval of confidence coefficient $(1 - \alpha)$, given by

$$\left\{ x_{(1)} - \frac{bS_r}{n(r-1)}, \quad x_{(1)} - \frac{aS_r}{n(r-1)} \right\}$$

However, there are several choices of (a, b) and again we must choose (a, b) by some definite rule. Usually 'equal tail' choice can be made, i.e., we take $a = F_{\alpha/2}(2, 2r-2)$ and $b = F_{1-\alpha/2}(2, 2r-2)$. However, we can choose (a, b) so that the length of the confidence interval $\frac{S_r}{n(r-1)} \cdot (b-a)$ is minimized, or equivalently $(b-a)$ is minimized subject to the condition $H_{2 \cdot 2r-2}(b) - H_{2 \cdot 2r-2}(a) = 1 - \alpha$. We note here that tables for such values of (a, b) are not available.

EXAMPLE 5.4   Consider the data for Example 1.3. Here $n = 30$, $r = 15$, $x_{(1)} = 650$ and $S_r = 14,000$. Suppose we want to test $H_0 : \mu \leqslant 450$ against $H_1 : \mu > 450$ at level $\alpha = 0.05$. By (15), the critical value for the statistic $x_{(1)}$ is given by $\mu_0 + \frac{S_r}{n(r-1)} F_{1-\alpha}(2, 2r-2)$. Substituting the appropriate values we have the critical value $= 450 + \frac{14,000}{30(14)} (3.35) = 561.67 \simeq 562$. We, therefore, reject H_0. We also remark here that $(-\infty, 538)$ gives us a 95% one-sided confidence interval for μ. If we want to obtain a 95% one-sided confidence interval for μ of the type (T_L, ∞), we use (18) and $T_L = x_{(1)} - \frac{S_r}{n(r-1)} F_\alpha(2, 2r-2)$. For the data, $T_L = 650 - \frac{14,000}{30(14)} \frac{1}{(19.45)} \simeq 648$. Thus, $(648, \infty)$ provides us with the required confidence interval. Here we have used the relationship $F_{1-\alpha}(m, n) = \frac{1}{F_\alpha(n, m)}$. The two-sided 90% confidence interval, using equal tail criteria, is given by $(538, 648)$.

If for the same data, we want to test at 5% level $H_0 : \sigma \leqslant 2000$ against $H_1 : \sigma \geqslant 2000$, then the rejection region is given by

$$S_r \geqslant \frac{\sigma_0}{2} \xi_{1-\alpha}(2r-2) = \frac{2000}{2} (41.337) = 41,337$$

Since the observed value of $S_r = 14,000$, we accept H_0. On the other hand if we want to test $H_0 : \sigma \geqslant 1500$ against $H_1 : \sigma < 1500$, then the critical point is given by

$$\frac{\sigma_0}{2} \xi_\alpha(2r-2) = \frac{1500}{2} (16.928) = 12,696$$

Here we reject H_0 if the observed value of $S_r < 12,696$. So here also we accept H_0.

The problem of testing hypotheses about μ and σ both is quite complicated and we will not consider these types of problems here. From practical consideration there are many such important problems, e.g., testing about $\psi(\mu, \sigma) = \mu + \sigma$, which represents expected life, or testing about $R_t(\mu, \sigma)$, the reliability function given by $\exp\left(-\dfrac{t - \mu}{\sigma}\right)$ if $t \geqslant \mu$ and 1 otherwise. However, here the solutions are very difficult to obtain and go beyond the scope of the present book (Ranganathan and Kale, 1978).

Normal Distribution

Let us consider the normal model. Let $(X_1, X_2, \ldots, X_n)$ be the failure times of n items put on the test and let $\{X_i\}_1^n$ be independent identically distributed normal r.v.'s with mean μ and variance σ^2. Thus, the likelihood of the sample is given by

$$L(x_1, \ldots, x_n \mid \mu, \sigma^2) = \left(\frac{1}{\sqrt{2\pi}}\right)^n \frac{1}{\sigma^n} \exp\left\{-\sum_{i=1}^n \frac{(x_i - \mu)^2}{2\sigma^2}\right\} \qquad (19)$$

Noting that $E(X) = \mu$, the mean life, we will first consider the test for $H_0 : \mu \geqslant \mu_0$ against $H_1 : \mu < \mu_0$. This corresponds to the case where a manufacturer wishes to test that the mean life of the product is at least μ_0. On the other hand if a buyer is purchasing the same product he would be interested in testing $H_0 : \mu \leqslant \mu_0$ against $H_1 : \mu > \mu_0$ since for a buyer the error in rejecting H_0, i.e., accepting $H_1 : \mu > \mu_0$ when in fact $\mu \leqslant \mu_0$, is more serious than accepting $H_0 : \mu \leqslant \mu_0$ when H_1 holds or $\mu > \mu_0$.

We will consider the problem from the manufacturer's view point and consider $H_0 : \mu \geqslant \mu_0$ against $H_1 : \mu < \mu_0$. Here σ^2 acts as a nuisance parameter. Let Ω_{H_0} denote the parameter space under H_0 and Ω the entire parameter space under the null as well as alternative hypotheses. The MLE's of (μ, σ^2) under H_0 are given by

$$\tilde{\mu} = \bar{x} \qquad \text{if } \bar{x} \geqslant \mu_0$$
$$= \mu_0 \qquad \text{if } \bar{x} < \mu_0$$
$$\tilde{\sigma}^2 = \frac{1}{n} \Sigma(x_i - \bar{x})^2 \qquad \text{if } \bar{x} \geqslant \mu_0$$
$$= \frac{1}{n} \sum_{i=1}^n (x_i - \mu_0)^2 \quad \text{if } \bar{x} < \mu_0$$

Therefore

$$\sup_{\Omega_{H_0}} \log L(x \mid \mu, \sigma^2) = -\frac{n}{2} \log 2\pi - \frac{n}{2} \log \tilde{\sigma}^2 - \frac{n}{2}$$

On the other hand

$$\sup_{\Omega} \log L(x \mid \mu, \sigma^2) = -\frac{n}{2} \log 2\pi - \frac{n}{2} \log \hat{\sigma}^2 - \frac{n}{2}$$

where

$$\hat{\sigma}^2 = \frac{1}{n} \sum_{i=1}^n (x_i - \bar{x})^2$$

This immediately gives the log of the likelihood ratio test statistics

$$\log \lambda(x) = \frac{n}{2} \log \hat{\sigma}^2 - \frac{n}{2} \log \tilde{\sigma}^2$$

Noting that $\hat{\sigma}^2 = \tilde{\sigma}^2$ for $\bar{x} \geqslant \mu_0$, we have

$$\log \lambda(x) = 0 \quad \text{if } \bar{x} \geqslant \mu_0$$

$$= \frac{n}{2} \log \sum_{i=1}^{n} (x_i - \bar{x})^2 \div \frac{n}{2} \log \sum_{i=1}^{n} (x_i - \mu_0)^2 \quad \text{if } \bar{x} < \mu_0$$

Now the rejection region consists of $\lambda(x) \leqslant c$ where $0 < c < 1$. Therefore, if $\bar{x} \geqslant \mu_0$, we always accept H_0. Now if $\bar{x} < \mu_0$, then we reject H_0 if

$$\lambda(x) = \left\{ \frac{\Sigma(x_i - \bar{x})^2}{\Sigma(x_i - \mu_0)^2} \right\}^{n/2} < c_1$$

or

$$\frac{\Sigma(x_i - \mu_0)^2}{\Sigma(x_i - \bar{x})^2} > \frac{1}{c_1}$$

Now $\quad \Sigma(x_i - \mu_0)^2 = \Sigma(x_i - \bar{x})^2 + n(\bar{x} - \mu_0)^2$

Hence we reject H_0 if

$$\frac{n(\bar{x} - \mu_0)^2}{\Sigma(x_i - \bar{x})^2} > \frac{1}{c_1} - 1 \quad \text{and} \quad \bar{x} < \mu_0$$

This is equivalent to rejecting H_0 if

$$t_{n-1} = \frac{\sqrt{n}(\bar{x} - \mu_0)}{\sqrt{\dfrac{1}{n-1} \Sigma(x_i - \bar{x})^2}} < c_2$$

where c_2 is determined such that

$$P\{t_{n-1} < c_2 \mid \mu, \sigma^2\} \leqslant \alpha \quad \text{for } \mu \geqslant \mu_0 \text{ and } \sigma^2 > 0 \qquad (20)$$

Again we select c_2 by using the boundary $\mu = \mu_0$ and $\sigma^2 > 0$ or

$$P\{t_{n-1} < c_2 \mid \mu_0, \sigma^2\} = \alpha, \ \sigma^2 > 0. \qquad (21)$$

Now the distribution of t_{n-1} under (μ_0, σ^2) is that of Student's-t with $(n-1)$ d.f. and $c_2 = t_{\alpha \cdot n-1}$ where $t_{p \cdot m}$ denotes the $100p\%$ point of a t-distribution with m degrees of freedom. Let s_x^2 denote $\dfrac{1}{n-1} \Sigma(x_i - \bar{x})^2$. Then we reject H_0 if

$$\bar{x} < \mu_0 + \frac{t_{\alpha \cdot n-1}}{\sqrt{n}} s_x \qquad (22)$$

We now show that if $\mu \geqslant \mu_0$, then

$$P\left\{ \bar{x} < \mu_0 + t_{\alpha \cdot n-1} \frac{s_x}{\sqrt{n}} \;\middle|\; \mu, \sigma^2 \right\} \leqslant \alpha$$

This can be seen as follows. When $\mu > \mu_0$,

$$\mu_0 + t_{\alpha \cdot n-1} \frac{s_x}{\sqrt{n}} < \mu + t_{\alpha \cdot n-1} \cdot \frac{s_x}{\sqrt{n}} \quad \text{for any } s_x^2$$

and therefore,

$$P\left\{\bar{x} < \mu_0 + t_{\alpha \cdot n-1} \frac{s_x}{\sqrt{n}} \Big| \mu, \sigma^2\right\} \leqslant P\left\{\bar{x} < \mu + t_{\alpha \cdot n-1} \frac{s_x}{\sqrt{n}} \Big| \mu, \sigma^2\right\}.$$

But

$$P\left\{\bar{x} < \mu + t_{\alpha \cdot n-1} \frac{s_x}{\sqrt{n}} \Big| \mu, \sigma^2\right\} = P\left\{\frac{\sqrt{n}(\bar{x} - \mu)}{s_x} < t_{\alpha \cdot n-1} \Big| \mu, \sigma^2\right\} = \alpha,$$

since under (μ, σ^2), $\sqrt{n}(x - \mu)/s_x$ is Student's-t with $(n - 1)$ degrees of freedom. Thus, the critical region given by (22) determines a size α test. It can be shown that this test has many desirable properties.

As a matter of fact, for obtaining confidence intervals we can use the pivotal quantity $z_{n-1} = \sqrt{n}(\bar{x} - \mu)/s_x$ which has Student's-t distribution with $(n - 1)$ degrees of freedom under any (μ, σ^2). Suppose we want to obtain a two-sided confidence interval of confidence coefficient $(1 - \alpha)$ of the type $T_L \leqslant \mu \leqslant T_U$. Then we consider Z_{n-1} and obtain constants (a, b) such that

$$P\{a \leqslant Z_{n-1} \leqslant b\} = G_{n-1}(b) - G_{n-1}(a) = 1 - \alpha \qquad (23)$$

where $G_{n-1}(u)$ is the distribution function of the Z_{n-1} which is Student's-t with $(n - 1)$ degrees of freedom under any (μ, σ^2). This immediately yields $T_L = \bar{x} - (bs_x/\sqrt{n})$ and $T_U = \bar{x} - (as_x/n)$. However, there are many values of (a, b) that satisfy (23). Again we determine (a, b) such that $(b - a)$ is minimized subject to (23). This immediately gives

$$b = t_{(1-\alpha/2) \cdot (n-1)} = -a$$

or the points (a, b) are determined by equal tail criteria. When n is large, we can use the normal approximation to t_{n-1} and $t_{(1-\alpha/2) \cdot (n-1)} = \xi_{1-\alpha/2}$ where ξ_p is the $100p\%$ point of the standard normal distribution.

To study tests of hypotheses about σ^2, one can show that these tests are based on the statistics

$$\Sigma(x_i - \bar{x})^2 = (n - 1) s_x^2 = S_x^2$$

and S_x^2/σ^2 is $\chi^2(n - 1)$. Thus, much of the work done for the testing of hypotheses about σ in one parameter exponential distribution can be directly used to obtain rejection regions and confidence intervals for σ^2 in the normal distribution model. Again as in the case of exponential distribution with two parameters, to obtain tests of hypotheses about functions of μ and σ is quite complicated and will not be attempted here. We refer to Lehmann (1959) for many problems of interest. The tests of hypotheses about the reliability function

$$R_t(\mu, \sigma) = 1 - \Phi\left(\frac{t - \mu}{\sigma}\right)$$

are in this category. Suppose we want to test $H_0 : R_t(\mu, \sigma) \geqslant p_0$ against $H_1 : R_t(\mu, \sigma) < p_0$ where t and p_0 are specified. If σ is known, then we can show that this is equivalent to testing $H_0 : \mu \geqslant t - \sigma\xi_{1-p_0} = \mu_0$ against

$H_1 : \mu < t - \sigma \xi_{1-p_0}$ and we can use the results proven earlier. However, if σ is unknown, then the problem is quite complex and we refer to Lehmann [(1959), pp. 222–224] and Resinikoff and Lieberman (1957).

Gamma and Weibull Distribution

Let us first consider the case of Gamma distribution with two parameters (σ, p) and the p.d.f. given by

$$f(x \mid \sigma, p) = \frac{1}{\Gamma(p)} \frac{1}{\sigma^p} \exp\left(-\frac{x}{\sigma}\right) x^{p-1}, \ x \geqslant 0, p > 0, \sigma > 0 \qquad (24)$$

Again we will consider the simplest cases only, namely testing for σ when p is known and testing for p when σ is known. We will not consider the problem of testing σ (or p) when the other parameter is unspecified.

Consider first the case when p is known. Then $Y_i = 2X_i/\sigma$ is a $\chi^2(2p)$ r.v. and $\sum_{i=1}^{n} Y_i = S$ is a $\chi^2(2np)$ r.v. Thus S is a pivotal quantity and we can construct confidence intervals for σ using S. The theory here is precisely the same as that for the exponential model with one unknown parameter σ. The only difference is that the pivotal quantity S now is $\chi^2(2np)$ rather than $\chi^2(2n)$. Similarly, the problems regarding tests of hypotheses about σ can be tackled with simple modifications. For example, if we want to construct one-sided confidence interval for σ with confidence coefficient $(1 - \alpha)$, then the lower and upper limits are given by

$$T_L = 2 \sum_{i=1}^{n} x_i / \xi_{1-\alpha}(2np) \text{ and } T_U = 2 \sum_{i=1}^{n} x_i / \xi_{\alpha}(2np)$$

Similarly, if we want to test $H_0 : \sigma \leqslant \sigma_0$ against $H_1 : \sigma > \sigma_0$, the critical region would be given by

$$\sum_{i=1}^{n} x_i \geqslant \frac{\sigma_0}{2} \xi_{1-\alpha}(2np)$$

where $\xi_{1-\alpha}(2np)$ is the $100(1 - \alpha)\%$ point of the χ^2 distribution with $2np$ degrees of freedom.

The case when σ is known and p is unknown is much more complicated. The basic difficulty is the fact that the MLE of p cannot be obtained explicitly and thus even the likelihood ratio test approach is quite difficult.

Since σ is known, let us take $\sigma = 1$, and consider the simplest problem of testing $H_0 : p = p_0$ against $H_1 : p = p_1$. By using the Neyman-Pearson lemma, we can immediately show that the critical region is

$$\sum_{i=1}^{n} \log x_i > c_1 \text{ if } p_1 > p_0$$

In case $p_1 < p_0$, the critical region becomes $\Sigma \log x_i < c_2$. Here the constants are to be determined by the size condition and these conditions are $F_{p_0}(c_1) = 1 - \alpha$ and $F_{p_0}(c_2) = \alpha$ respectively where $F_{p_0}(u)$ denotes the distribution function of $\Sigma \log x_i$ under $p = p_0$. Unfortunately, this distribution is not tractable. Therefore, exact tests which are optimum are difficult to obtain. Approximations to the most powerful tests can be obtained when np_0 is large.

It may be noted here that the moment estimator of p is given by $(1/n) \Sigma x_i$ and $S = 2\Sigma x_i$ is $\chi^2(2np)$. Thus, we can obtain tests or confidence intervals for p using $S = 2\Sigma x_i$, rather than $T = \sum_{i=1}^{n} \log x_i$. Of course, the tests based on S would not be as good as those based on T since T is sufficient for p while S is not. However, we can obtain only approximations to the tests based on T while we can obtain exact tests based on S. Thus, for testing $H_0 : p = p_0$ against $H_1 : p_1 > p_0$ (say), we consider the r.v. S which under H_0 is $\chi^2(2np_0)$ and under H_1 is $\chi^2(2np_1)$ distribution. We therefore apply Neyman-Pearson lemma to S and then obtain the critical region $S > \xi_{1-\alpha}(2np_0)$. This is an exact level α test. Further it is uniformly more powerful for $H_1 : p > p_0$ within the class of tests based on S.

One can also construct confidence intervals for p using the relation S is $\chi^2(2np)$. This was done by Bain and Weeks (1965). In particular for one-sided confidence interval of the type $(0, T_U)$ of confidence coefficient $(1 - \alpha)$, T_U is the solution of $S = \xi_{1-\alpha}(2nT_U)$ which can be obtained by using tables of χ^2 distribution. Thus, suppose that the observed value of S is 40 and $1 - \alpha = .95\%$ and $n = 15$, using tables of $\chi^2(30 \cdot p)$ for various values of p, we can obtain $T_U = 1.874$.

Next, we consider the Weibull distribution where the p.d.f. is given by

$$f(x \mid \sigma, p) = \frac{p}{\sigma} \frac{x^{p-1}}{\sigma} \exp\left\{-\left(\frac{x}{\sigma}\right)^p\right\}, \quad x > 0, p > 0, \sigma > 0 \qquad (25)$$

As observed earlier, if p is known, then by using the transformation $Y = X^p$, we can reduce the problem to the exponential density with one parameter and the result obtained earlier can be directly used.

The general theory of obtaining tests of hypotheses about p when σ is known or unknown is quite complicated and we give only the basic results and refer the interested reader to the source for details.

Thus, for the case where σ is known and p unknown, Bain and Weeks (1965) obtained the confidence limits for p which can be used to obtain tests of hypotheses of the type $H_0 : p = p_0$ against $H_1 : p > p_0$. The more interesting case when p and σ are both unknown has been treated by Thoman, Bain and Antle (1969) and Bain and Weeks (1965). Based on certain functions of the maximum likelihood estimators $\hat{p}$ and $\hat{\sigma}$, Thoman, Bain and Antle (1969) obtained confidence intervals for each parameter when the other is unspecified. This paper tabulates the percentage points of such functions for a wide range of values of α and different sample sizes n. The paper also plots the power function of the test $H_0 : p = p_0$ against $H_1 : p = p_1$ for various combinations of (n, p_1). We illustrate by an example the use of techniques developed in these papers. The tables and figures referred to in this example refer to Thoman, Bain and Antle (1969).

EXAMPLE 5.5   Consider the data in the Example 2.2. Suppose we wish to test $H_0 : p = 1$ against $H_1 : p \neq 1$. This problem is of interest in that for $p = 1$ we get the exponential distribution with mean life σ. Here the MLE of $\hat{p} = 2.035$. We now use the statistics $\hat{p}/p$ and using Table 1 determine confidence intervals for p of the confidence coefficient $1 - \alpha = .90$ say. Thus, we obtain $p \cdot l_\alpha$ points for $\alpha = .05$ and $\alpha = .95$ where l_α is such that

$$P\left\{\frac{\hat{p}}{p} \leqslant l_\alpha \,|\, p, \sigma\right\} = \alpha$$

These are given by $l_{.05} = 0.8075$ and $l_{.95} = 1.381$. Therefore, we have

$$P\left\{\frac{\hat{p}}{l_{.95}} < p < \frac{\hat{p}}{l_{.05}} \,|\, p, \sigma\right\} = .90$$

Here $\dfrac{\hat{p}}{l_{.95}} = 1.474$ and $\dfrac{\hat{p}}{l_{.05}} = 2.520$ and the 90% confidence interval for this data is given by (1.474, 2.520). Since the value under $H_0 : p = 1$ lies outside the confidence interval, we reject H_0 at the level $\alpha = .10$. We note that if the problem is to test $H_0 : p = 1$ against $H_1 : p > 1$, then we obtain one-sided confidence interval $\left(\dfrac{\hat{p}}{l_{1-\alpha}}, \infty\right)$ and if $\dfrac{\hat{p}}{l_{1-\alpha}} > 1$, we reject H_0 at level α. Thus, for the above data at $\alpha = .05$, we have $\dfrac{\hat{p}}{l_{.95}} = 1.474$ and we reject H_0 here also. Using the figure 1a, the power of this test at $p = 1.5$ is about .85.

For the same data suppose we want to test $H_0 : \sigma = 2$ against $H_1 : \sigma \neq 2$ at level $\alpha = .10$. Then we obtain the two-sided confidence interval for σ of confidence coefficient $1 - \alpha = .90$ by using the percentage points of $\hat{p} \log (\hat{\sigma}/\sigma)$. Thus, we obtain $l_{.05}$ and $l_{.95}$ where l_α is given by

$$P\left\{\hat{p} \log \frac{\hat{\sigma}}{\sigma} \leqslant l_\alpha\right\} = \alpha$$

We have

$$P\left\{l_{.05} \leqslant \hat{p} \log \frac{\hat{\sigma}}{\sigma} \leqslant l_{.95} \,|\, p, \sigma\right\} = 0.90$$

and a simple algebraic manipulation gives

$$\left\{\hat{\sigma} \exp\left(-\frac{l_{.95}}{\hat{p}}\right), \; \hat{\sigma} \exp\left(-\frac{l_{.05}}{\hat{p}}\right)\right\}$$

as the two-sided 90% confidence interval for σ. For the data the upper and lower limits are given by 1.517 and 2.188 respectively and we accept H_0 since the value under $H_0 : \sigma = 2$ lies inside the observed confidence interval.

Suppose we want to test $H_0 : \sigma \geqslant 2$ against $H_1 : \sigma < 2$ at $\alpha = 0.10$. We obtain a one-sided 90% confidence interval for σ of the type (σ_L, ∞) where σ_L is chosen such that $P\{\sigma_L < \sigma \,|\, p, \sigma\} = 1 - \alpha$. Again using

$$P\left\{\hat{p} \log \frac{\hat{\sigma}}{\sigma} \leqslant l_\alpha \,|\, p, \sigma\right\} = \alpha$$

we have

$$0.90 = P\left\{\hat{p} \log \frac{\hat{\sigma}}{\sigma} \leqslant l_{.90} \mid p, \sigma\right\}$$

which immediately leads to

$$\sigma_L = \hat{\sigma} \exp \left(\frac{l_{.90}}{\hat{p}}\right) = 1.5842$$

for the data under consideration and we do not reject H_0.

The problem of the testing of reliability function $R_{t_0}(p, \sigma) = R_0$ against $H_1 : R_{t_0} > R_0$ at level α has been considered by Johns Jr. and Liebermann (1966). This paper gives the one-sided confidence interval of the type $(R_L(t_0), 1)$ for usual values of $1 - \alpha$. If the specified $R_0 < R_L(t_0)$, then we reject H_0; otherwise we accept it.

EXERCISES

5.1   100 electronic tubes of a certain type were tested and the test was terminated after the first 15 tubes failed. The failure times (in hours) are given by: 40, 60, 90, 120, 195, 260, 350, 420, 501, 620, 650, 730, 815, 910, 980. Assuming that the failure time distribution is exponential with mean life σ, test the following hypotheses at $\alpha = .10$ about the reliability function $R_{t_0}(\sigma)$ at $t_0 = 800$ hours.

$$H_0 : R_{t_0}(\sigma) \geqslant .95 \text{ against } H_1 : R_{t_0}(\sigma) < .95$$

Obtain also the one-sided confidence intervals of confidence coefficient 0.90 for $R_{t_0}(\sigma)$.

5.2   40 items were placed on test and the test was terminated after 16 failures. The first failure occurred at 525 hours after the test started and the observed total time between the first and the 16th failure was 15,000. Assuming that the life distribution follows two-parameter exponential density, test the following hypotheses at $\alpha = 0.01$.
(a) $H_0 : \mu \leqslant 600$ against $H_1 : \mu > 600$, σ unspecified
(b) $H_0 : \sigma \geqslant 1500$ against $H_1 : \sigma < 1500$, μ unspecified
Obtain the two-sided 90% confidence interval for σ for the above data. Also obtain the 90% confidence intervals for μ of the type $(-\infty, \mu_U)$ and (μ_L, ∞).

5.3   15 units were subjected to test and the test continued until all items failed. The mean failure time was 10 hours. If the failure time distribution is given by

$$f(x, \sigma) = \frac{2}{\sqrt{\pi \sigma}} \frac{\sqrt{x}}{\sigma} \exp\left\{-\frac{x}{\sigma}\right\}, x > 0, \sigma > 0,$$

test the hypothesis that $\sigma = 5$ against $H_1 : \sigma > 5$ at $\alpha = 0.10$.

5.4   10 items were tested giving the mean failure time 16 hours. If the failure time distribution follows the gamma p.d.f. with $\sigma = 10$ and p unknown, test the hypothesis $H_0 : p = 1$ against $H_1 : p > 1$ at $\alpha = 0.05$.

5.5   Consider the data given in Exercise 2.5. The failure time distribution is $f(x, \sigma, p) = \frac{2x}{\sigma^2} \exp\left\{-\frac{x^2}{\sigma^2}\right\}, x > 0, \sigma > 0$. Using the transformation $y = x^2$, test the hypotheses about $H_0 : \sigma \leqslant 10$ against $H_1 : \sigma > 10$.

5.6   Consider the data of Exercise 2.8. Test the following hypotheses at $\alpha = 0.05$.
(a) $H_0 : b = 1.5$ against $H_1 : b > 1.5$
(b) $H_0 : c = 4$ against $H_1 : c < 4$.

5.7 Suppose n items are put to test and it was decided to truncate the experiment after r failures. Suppose that the failure time distribution is a single parameter exponential with mean life σ. It is decided to test $H_0 : \sigma = 2$ against $H_1 : \sigma > 2$ at level $\alpha = 0.01$ and such that the power of the test at $\sigma = 5$ is 0.90. Determine the smallest value of r such that the test satisfies these specifications. Also compute the power of this test at $\sigma = 3$, 4 and 6.

5.8 Let the failure time distribution be $N(\mu, \sigma^2)$. The 90% confidence interval is given by $\bar{x} \pm t_{.95}(n-1) \dfrac{s_x}{\sqrt{n}}$, where $t_\alpha(n-1)$ denotes the $100\alpha\%$ point of the t distribution with $(n-1)$ degrees of freedom. Determine n such that the expected length of the confidence interval is less than $\sigma/2$. Note that $(n-1)s_x^2/\sigma^2$ is $\chi^2(n-1)$ and

$$E(s_x) = \sigma \sqrt{\frac{2}{n-2}} \; \frac{\Gamma\left(\frac{n}{2}\right)}{\Gamma\left(\frac{n-1}{2}\right)}.$$

CHAPTER 6
Bayes Estimators

The framework for the results in the earlier chapters so far has been as follows. We have a random sample of size n, say $(X_1, X_2, \ldots, X_n)$ which we regard as independent identically distributed random variables with distribution function (d.f.) $F(x, \theta)$ and p.d.f. $f(x, \theta)$, where θ is a labelling parameter, real valued or vector valued as the case may be. Quite often this parameter θ represents mean life time, or failure rate or other such quantities associated with the underlying population from which the sample has been obtained. We also assumed that we do not know the exact value of the parameter θ. However we agreed upon the parameter space, i.e. the set of all possible values of the parameter which we denote by Ω. Thus, in the exponential case we assumed that the p.d.f. of the failure time is given by $f(x, \theta) = \dfrac{1}{\theta} \exp\left(-\dfrac{x}{\theta}\right)$ where $x > 0$ and $\theta > 0$. Here $\Omega = \{\theta \mid \theta > 0\}$ is the parameter space. Mostly we have assumed that Ω is the natural parameter space, i.e., Ω contains those values of the parameter θ for which the p.d.f. $f(x, \theta)$ is well defined, i.e., $f(x, \theta) \geqslant 0$ and $\displaystyle\int_{-\infty}^{\infty} f(x, \theta)\, dx = 1$. Similarly, in the case of mixtures of exponentials we assumed that

$$f(x, \theta) = p\, \frac{1}{\theta_1} \exp\left(-\frac{x}{\theta_1}\right) + (1 - p)\, \frac{1}{\theta_2} \exp\left(-\frac{|x}{\theta_2}\right), \quad x > 0$$

and

$$\Omega = \{(p, \theta_1, \theta_2) \mid 0 \leqslant p \leqslant 1,\ \theta_1 > 0,\ \theta_2 > 0\}$$

There are cases in which one can assume a little more about the unknown parameter $\theta \in \Omega$. For example, we could assume the θ is itself a random variable with d.f. $F(\theta)$ or p.d.f. $f(\theta)$. For example, in the exponential model the mean life θ may be regarded as varying from batch to batch over time and this variation is represented by a probability distribution over Ω. Thus, the set up is now viewed as follows.

There are n items put to test and it is assumed that their recorded life times form a random sample of size n from a population with p.d.f. $f(x, \theta)$. To be specific we will assume θ to be real valued. We agree to regard θ itself as a random variable with a p.d.f. $g(\theta)$. Thus, the failure time p.d.f. $f(x, \theta)$ is regarded as a conditional p.d.f. $f(x \mid \theta)$ where the marginal p.d.f. of θ is given by $g(\theta)$. Thus, the joint p.d.f. of $(X_1, X_2, \ldots, X_n, \theta)$ is given by

$$H(x_1, \ldots, x_n, \theta) = \prod_{i=1}^{n} f(x_i \mid \theta)g(\theta) = L(x_1, \ldots, x_n \mid \theta)g(\theta) \qquad (1)$$

the marginal p.d.f. of $(X_1, X_2, \ldots, X_n)$ is given by

$$P(X_1, \ldots, X_n) = \int_{\Omega} H(x_1, \ldots, x_n, \theta) \, d\theta \qquad (2)$$

and the conditional p.d.f. of θ given the data $(x_1, \ldots, x_n)$ is given by

$$\Pi(\theta \mid x_1, \ldots, x_n) = \frac{H(x_1, \ldots, x_n, \theta)}{P(x_1, \ldots, x_n)} = \frac{L(x_1, x_2, \ldots, x_n \mid \theta)g(\theta)}{\displaystyle\int_{\Omega} L(x_1, x_2, \ldots, x_n \mid \theta)g(\theta) d\theta}$$
$$(3)$$

Thus, prior to obtaining the data $(x_1, x_2, \ldots, x_n)$, the variations in θ were represented by $g(\theta)$, known as the prior distribution on Ω; however, after the data $(x_1, x_2, \ldots, x_n)$ has been obtained, in the light of the new information, the variations in θ are represented by $\Pi(\theta \mid x_1, \ldots, x_n)$, the posterior distribution of θ. The uncertainty about the parameter θ prior to experiment is represented by the prior p.d.f. $g(\theta)$ and the same after the experiment is represented by the posterior p.d.f. $\Pi(\theta \mid x_1, \ldots, x_n)$. This process is a straightforward application of Bayes theorem.

Once the posterior distribution has been obtained it becomes the main object of study and it is conventional to take as Bayes estimator of θ the mean of the posterior distribution $\Pi(\theta \mid x_1, \ldots, x_n)$. Thus, the Bayes estimator of θ in the above set up is given by

$$E(\theta \mid x_1, x_2, \ldots, x_n) = \int_{\Omega} \theta \Pi(x_1, x_2, \ldots, x_n \mid \theta) \, d\theta \qquad (4)$$

As we shall see in later parts of this chapter, many estimators derived in the previous chapters are Bayes or nearly the same as Bayes estimators for specific priors assumed on the parameter space Ω.

The approach given above is known as Bayesian after Rev. Thomas Bayes (1763) whose first paper outlined the basic theory. Some of the standard texts giving more detailed discussion are those of Lindley (1965) and Box and Tiao (1973). For the basic theory and foundations one can refer to Savage (1962) and Jeffreys (1961). Ever since the original scheme was proposed by Bayes a crucial problem has been the prior distribution $g(\theta)$. How does one select a known prior distribution $g(\theta)$ to express the uncertainty about the unknown parameter θ?

There may be some empirical evidence obtained through earlier experiments which would help us decide on the prior distribution $g(\theta)$. On the other hand one can decide on $g(\theta)$, or at least a class of priors $g(\theta)$, on a subjective basis or in a normative way. Various rules have been suggested and it appears that there is no neat solution to the problem.

Raifa and Schlaifer (1961) have introduced what are known as 'conjugate priors' which contain a variety of prior distributions which are comparatively easy to handle. A general rule proposed by Jeffreys prescribes $g(\theta) = $ constant $\sqrt{I(\theta)}$ where $I(\theta)$ is the well known Fisher information defined by

$$I(\theta) = E\left[\left(\frac{\partial \log f(x, \theta)}{\partial \theta}\right)^2\right] = E\left(\frac{-\partial^2 \log f(x, \theta)}{\partial \theta^2}\right) \qquad (5)$$

when there is only one unknown parameter θ. For vector valued parameter we can use $|I(\theta)|$, the determinant of the information matrix $I(\theta) = E\left(\frac{-\partial^2 \log f}{\partial \theta_i \, \partial \theta_j}\right)$. Another important concept that crops up here is the use of improper priors or quasi-priors. Such priors arise when $g(\theta)$ is not a probability distribution in that $g(\theta) \geqslant 0$ but $\int_\Omega g(\theta) \, d\theta$ is not equal to one. Without going into many details about these controversial issues, we will study the behaviour of the Bayes estimators under proper and improper priors for three well known and widely used failure time distributions—Exponential, Normal and Weibull.

Exponential Distribution

Consider the one-parameter exponential failure time distribution

$$f(x \mid \theta) = \frac{1}{\theta} \exp\left(-\frac{x}{\theta}\right), \; x > 0, \quad \theta > 0$$

where θ is now treated as a random variable. A straightforward computation gives the Fisher information

$$I(\theta) = E\left\{-\frac{\partial^2}{\partial \theta^2} \log f(x \mid \theta)\right\} = \frac{1}{\theta^2}$$

Hence Jeffreys prior $g(\theta) = 1/\theta$, which is an improper (or quasi) prior since $\int_0^\infty g(\theta) \, d\theta$ is not finite.

Let us consider a more general class of priors, $g(\theta) \propto (1/\theta^c)$, $c > 0$. We have $L(\theta \mid \mathbf{x}) = 1/\theta^n \exp\left(-\sum_1 x_i/\theta\right)$, where $\mathbf{x} = (x_1, x_2, \ldots, x_n)$. The posterior distribution of θ is given by

$$\Pi(\theta \mid \mathbf{x}) = \frac{K}{\theta^{n+c}} \exp\left(\frac{-n\bar{x}}{\theta}\right)$$

where

$$K^{-1} = \int_0^\infty \frac{1}{\theta^{n+c}} \exp\left(\frac{-n\bar{x}}{\theta}\right) d\theta$$

We will frequently use the result

$$\int_0^\infty \frac{\exp\left(-a/x\right)}{x^{m+1}} \, dx = \frac{\Gamma(m)}{a^m}$$

Substituting $a = n\bar{x}$, $m = (n + c - 1)$ we have

$$K^{-1} = \frac{\Gamma(n + c - 1)}{(n\bar{x})^{n+c-1}}$$

Thus,

$$\Pi(\theta \mid x) = \frac{(n\overline{x})^{n+c-1} \exp\left(\frac{-n\overline{x}}{\theta}\right)}{\Gamma(n+c-1)\theta^{n+c}}, \quad \theta > 0 \qquad (6)$$

and the Bayes estimator of $\theta = E(\theta \mid x) = \int_0^\infty \theta \Pi(\theta \mid x)\, d\theta = \theta^*$ (say).

From (6), by direct integration we obtain

$$\theta^* = \frac{(n\overline{x})^{n+c-1}}{\Gamma(n+c-1)} \int_0^\infty \frac{\exp\left(\frac{-n\overline{x}}{\theta}\right)}{\theta^{n+c-1}}\, d\theta$$

$$= \frac{(n\overline{x})^{n+c-1}\Gamma(n+c-2)}{\Gamma(n+c-1)(n\overline{x})^{n+c-2}}$$

$$= \frac{n\overline{x}}{n+c-2}$$

For $c = 1$, $g(\theta) = 1/\theta$ (Jeffreys prior), we have

$$\theta_1^* = \frac{n\overline{x}}{n-1}$$

For $c = 2$, $g(\theta) = 1/\theta^2$, and we get $\theta_2^* = \overline{x}$ which is the maximum likelihood as well as the uniformly minimum variance unbiased estimator (UMVUE) of θ.

For $c = 3$, $g(\theta) = 1/\theta^3$ and $\theta_3^* = n\overline{x}/(n+1)$, the well-known minimum—mean square-error (MSE) estimator of θ, in that θ_3^* has the smallest MSE in the class of linear estimators of the type $\Sigma l_i x_i$.

We note that for n quite large as compared to c, θ_1^*, θ_2^*, θ_3^* will all be numerically very close to each other. Thus, the effect of the prior distribution on actual estimators is rather small when the sample size is large. This fact is well exhibited in the equation (3), namely

$$\Pi(\theta \mid x) = g(\theta) \cdot L(x_1, \ldots, x_n \mid \theta)$$

which shows that at least in large samples the posterior is more dominated by the likelihood $\prod_{i=1}^n f(x_i \mid \theta) = L(x_1, \ldots, x_n \mid \theta)$ than the prior $g(\theta)$. Moreover, we can show that $E[(\theta_{c_1}^* - \theta_{c_2}^*)^2] \to 0$ as $n \to \infty$ for any fixed c_1 and c_2. By Tchebychev's inequality it follows that for any $\epsilon > 0$ $\lim_{n \to \infty} P\{|\theta_{c_1}^* - \theta_{c_2}^*| < \epsilon\} = 1$, showing thereby that in large samples the choice of the constant c (i.e., the choice of the prior distribution) is not very crucial.

Consider another class of priors given by

$$g(\theta) \propto \frac{\exp(-a/\theta)}{\theta^c}, \quad c \geqslant 0, a \geqslant 0, \theta > 0$$

$g(\theta)$ is a proper prior for $c > 1$. The posterior of θ is given by

$$\Pi(\theta \mid x) = \frac{(n\overline{x} + a)^{n+c-1}}{\Gamma(n+c-1)\theta^{n+c}} \exp\left(-\frac{n\overline{x}+a}{\theta}\right)$$

and Bayes estimator is given by

$$\theta^* = \frac{n\bar{x} + a}{n + c - 2}$$

Note that by putting $a = 0$, we have the results obtained earlier.

EXAMPLE 6.1   The following is a sample of 20 observations from an exponential p.d.f. (3) with $\theta = 4$.

SAMPLE

3.4177	3.5351	3.2259	2.7425	1.7323
2.6155	2.4651	2.3467	2.9487	0.5937
0.5625	3.1356	0.3484	3.7287	3.5680
9.3208	1.8474	0.3994	9.2817	17.2271

Table 6.1   Bayes estimates of θ

c \\ a	0	1	2	3	$\left(\dfrac{\text{Min.}}{\text{Max.}} \middle/ c\right)$
0	4.1690	4.2246	4.2802	4.3357	0.9616
1	3.9496	4.0022	4.0549	4.1075	0.9616
2	3.7521	3.8021	3.8521	3.9021	0.9616
3	3.5735	3.6211	3.6687	3.7163	0.9616
$\left(\dfrac{\text{Min.}}{\text{Max.}} \middle/ a\right)$	0.8572	0.8572	0.8572	0.8572	

Note that (Min./Max.) for fixed c and varying a is $n\bar{x}/(n\bar{x} + a)$ and for large n would be very close to 1. Similarly (Min./Max.) for fixed a and varying c is $(n - 2)/(n + 1)$ and for large n would again be very near to one—showing thereby that for large samples, choice of prior distribution is not crucial. Further, we note that the Bayes estimators are more robust to changes in a than to those in c. We now consider an inverted gamma (Raiffa and Schlaifer, 1961) as the prior distribution of θ. Such a prior is given by

$$g(\theta) = \frac{1}{a\Gamma(b)} \exp\left(-\frac{a}{\theta}\right)\left(\frac{a}{\theta}\right)^{b+1}, \quad 0 < \theta < \infty, \ a, b > 0 \qquad (7)$$

Note that (7) is a proper prior. The posterior distribution of θ may be obtained by straightforward integration and is given by

$$\Pi(\theta \mid x) = \frac{(n\bar{x} + a)^{n+b} \exp\left(-\dfrac{n\bar{x} + a}{\theta}\right)}{\theta^{n+b+1}\Gamma(n + b)}, \quad \theta > 0, \ a, b > 0$$

and the corresponding Bayes estimator

$$\theta^* = \frac{n\bar{x} + a}{n + b - 1}$$

This is the same as the estimator considered before with $b = c - 1$. Note that the difference between θ^* and the MLE $\hat{\theta} = \bar{x}$ is numerically small when n is large in relation to a and b. Indeed the difference $|\theta^* - \hat{\theta}|$ converges in probability to zero as $n \to \infty$.

If instead of a complete sample we have a failure censored sample (after $r \leqslant n$ failures), much of the theory remains the same and the Bayes estimators are obtained by replacing $n\bar{x} = \sum_{i=1}^{n} x_i$ by $S_r = \sum_{i=1}^{n} x_{(i)} + (n-r)x_{(r)}$ and n by r. Here $\{x_{(1)} < x_{(2)} < \ldots < x_{(r)}\}$ are the order statistics of the censored sample.

We will now consider two-parameter exponential density with p.d.f.

$$f(x \mid \mu, \theta) = \frac{1}{\theta} \exp \left\{ \frac{-(x - \mu)}{\theta} \right\}, \quad -\infty < \mu < x < \infty, \ \theta > 0$$

We consider the case of the failure censored sample. Let $x_{(1)} < x_{(2)} < \ldots < x_{(r)}$ be the first r failure times. Then as per results in Chapter 1 the likelihood is given by

$$L(x \mid \mu, \theta) = \frac{n!}{(n-r)! \ \theta^r} \exp \left\{ - \frac{\sum_{i=1}^{n} (x_i - \mu) + (n-r)(x_{(r)} - \mu)}{\theta} \right\}$$

$$= \frac{K}{\theta^r} \exp \left[-\frac{1}{\theta} \{S + n(x_{(1)} - \mu)\} \right]$$

where $S = \sum_{i=1}^{r} (x_{(i)} - x_{(1)}) + (n-r)(x_{(r)} - x_{(1)})$ and K is a normalizing constant. We consider a class of quasi-prior distributions $g(\mu, \theta) \propto \frac{1}{\theta^a}$, $-\infty < \mu < \infty$, $a, \theta > 0$.

The posterior distribution of (μ, θ) is given by

$$\Pi(\theta, \mu \mid x) = \frac{K}{\theta^{r+a}} \exp \left[-\frac{1}{\theta} \{S + nx_{(1)} - n\mu\} \right]$$

where

$$x = \{x_{(1)}, x_{(2)}, \ldots, x_{(r)}\} \text{ and } K^{-1} = \int_{-\infty}^{x_{(1)}} \int_{0}^{\infty} \Pi(\theta, \mu \mid x) \, d\theta \, d\mu$$

$$= \frac{\Gamma(r + a - 2)}{nS^{r+a-2}}$$

The marginal posterior of μ

$$\Pi(\mu \mid x) = \int_{0}^{\infty} \Pi(\mu, \theta \mid x) \, d\theta$$

$$= \frac{(r + a - 2)nS^{r+a-2}}{\{S + nx_{(1)} - n\mu\}^{r+a-1}}, \quad -\infty < \mu < x_{(1)}$$

Bayes estimator is given by

$$\mu^* = n(r + a - 2)S^{r+a-2} \int_{-\infty}^{x_{(1)}} \frac{\mu \, d\mu}{\{S + nx_{(1)} - n\mu\}^{r+a-1}}$$

$$= x_{(1)} - \frac{S}{n(r + a - 3)}, \quad r + a > 3 \tag{8}$$

Similarly, the marginal posterior of θ

$$\Pi(\theta \mid x) = \int_{-\infty}^{x_{(1)}} \Pi(\theta, \mu \mid x) \, d\mu$$

$$= \frac{S^{r+a-2}}{\Gamma(r + a - 2)} \left\{ \frac{\exp(-S/\theta)}{\theta^{r+a-1}} \right\}, \quad \theta > 0$$

and Bayes estimator is given by

$$\theta^* = \frac{S^{r+a-2}}{\Gamma(r + a - 2)} \int_0^\infty \frac{\exp(-S/\theta) \, d\theta}{\theta^{r+a-2}}$$

$$= \frac{S}{r + a - 3} \tag{9}$$

For 'a complete sample' case $(r = n)$ and $a = 2$, from (8) and (9) Bayes estimators are

$$\mu^* = \frac{nx_{(1)} - \overline{x}}{n - 1}, \qquad \theta^* = \frac{n\{x_{(1)} - \overline{x}\}}{n - 1}$$

which are identical with the UMVUE of (μ, θ).

Using Grubbs' (1971) data the posteriors of μ and θ are plotted in Figures 1 and 2. The posteriors of μ are quite robust for varying a in the quasi-prior $p(\mu, \theta) \propto (1/\theta^a)$, while the posteriors of θ are less so.

Next, we consider Bayes Estimator of the reliability function $R_t = P\{X \geqslant t\} = \exp\{-t/\theta\}$ for the single parameter exponential distribution with mean life θ. Again we will consider the failure censored samples only. Let $\{x_{(1)} < x_{(2)} < \ldots < x_{(r)}\}$ be the first r failure times. Assuming quasi prior $g(\theta) \propto (1/\theta)$, we have the posterior distribution of θ as given by

$$\Pi(\theta \mid x) = \frac{S_r^r}{\Gamma(r)} \theta^{-(r+1)} \exp\left(\frac{-S_r}{\theta}\right) \tag{10}$$

where

$$S_r = \sum_{i=1}^r x_{(i)} + (n - r)x_{(r)}$$

Note that (10) is a special case of (6) with $r = n, c = 1$.

Bayes estimator of R_t is given by

$$R_t^* = E(R_t \mid x)$$

$$= \int_0^\infty \exp(-t/\theta) \, \Pi(\theta \mid x) \, d\theta$$

$$= \frac{1}{\left(1 + \dfrac{t}{S_r}\right)^r} \tag{11}$$

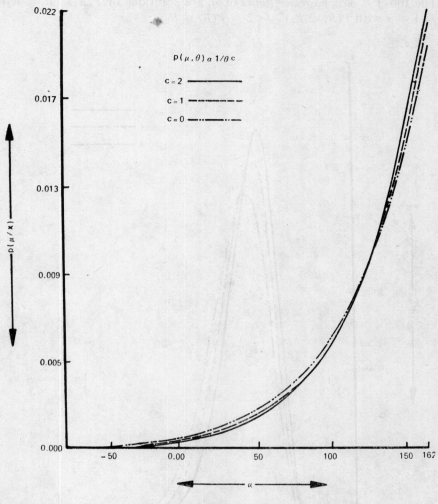

Fig. 1   Posterior of μ

This result is due to Bhattacharya (1967).

Note that whereas the uniformly minimum variance unbiased estimator of reliability

$$\tilde{R}_t = \left(1 - \frac{t}{S_r}\right)^{r-1}, \ t \leqslant S_r$$

and

$$= 0, \ t > S_r$$

the Bayes estimator

$$R_t^* = \frac{1}{(1 + t/S_r)^r} \text{ is non-zero for all } t > 0$$

The $100\,(1-\alpha)\%$ *posterior limits* of R_t are solutions of $P(R_1 < R_t < R_2)$ $= 1 - \alpha$ with $P(R_t < R_1) = \alpha/2 = P(R_t > R_2)$.

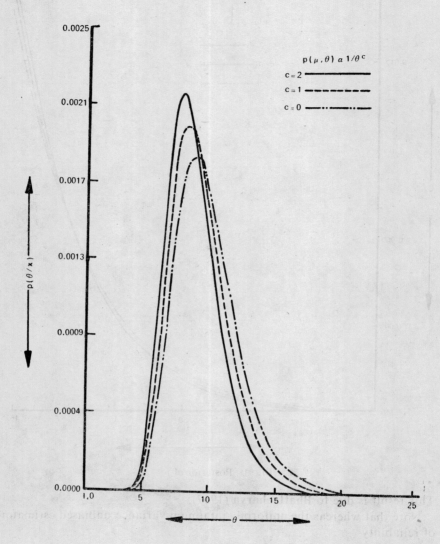

Fig. 2   Posterior of θ

$$\frac{\alpha}{2} = P(R_t < R_1) = P\{\exp\,(-t/\theta) < R_1\}$$

$$= P\left(\frac{t}{\theta} > \log\frac{1}{R_1}\right) = P\left(\theta < \frac{t}{\log\dfrac{1}{R_1}} = \theta_1\right)$$

Thus $R_1 = \exp(-t/\theta_1)$ where θ_1 satisfies

$$\frac{\alpha}{2} = \int_0^{\theta_1} \Pi(\theta \mid x) \, d\theta = \frac{1}{\Gamma(r)} \int_0^{\theta_1} \frac{S_r^r \exp(-S_r/\theta)}{\theta^{r+1}} \, d\theta$$

$$= \frac{1}{\Gamma(r)} \int_{S_r/\theta_1}^{\infty} \exp\left(-\frac{\chi^2}{2}\right)\left(\frac{\chi^2}{2}\right)^{r-1} d\left(\frac{\chi^2}{2}\right)$$

Hence

$$\frac{2S_r}{\theta_1} \sim \chi^2(2r), \quad \theta_1 = \frac{2S_r}{\chi^2_{1-\alpha/2}}$$

Similarly

$$R_2 = \exp(-t/\theta_2), \quad \theta_2 = \frac{2S_r}{\chi^2_{\alpha/2}(2r)}$$

where $\chi^2_p(m) = 100p\%$ point of a χ^2 distribution with m degrees of freedom. From the sampling distribution of the MLE $\hat{\theta}_r = S_r/r$ we know that $100(1-\alpha)\%$ confidence limits of θ are $\left\{\dfrac{2S_r}{\chi^2_{1-\alpha/2}(2r)}, \dfrac{2S_r}{\chi^2_{\alpha/2}(2r)}\right\} \equiv (\theta_1, \theta_2)$ and therefore, the corresponding *confidence limits* of R_t are $\left\{\exp\left(\dfrac{-t}{\theta_1}\right), \exp\left(\dfrac{-t}{\theta_2}\right)\right\}$.

It follows that Bayes posterior bounds and confidence limits are identical. This situation is analogous to the situation where Bayes estimator for some prior turns out to be the same as the MLE or UMVUE.

However, there are cases in which these limits are not the same for any prior. For further details we refer to Barnett (1973). Using the transformation $\theta = -t/\log R_t$ one can directly obtain the posterior distribution of R_t from (10), viz.,

$$\Pi(R_t \mid x) = \left(\frac{S_r}{t}\right)^r \frac{1}{\Gamma(r)} (-\log R_t)^{r-1}(R_t)^{(S_r/t)-1}, \quad 0 < R_t < 1$$

EXAMPLE 6.2 Consider the data of Example 1.2. Let us compute Bayes estimate of R_t and set 95% posterior bounds $R_1(t)$, $R_2(t)$ at $t = 200$ hours.

We have $n = 40, r = 10, S_{10} = \sum_n^{10} x_{(i)} + 50x_{10} = 50{,}676$

From (11), the Bayes estimate is given by

$$R^*(200) = \left(1 + \frac{200}{50{,}676}\right)^{-10} = 0.9614$$

The values of χ^2 are

$$\chi^2_{0.975}(20) = 34.1696, \quad \chi^2_{0.025}(20) = 0.5908$$

Thus, $(0.9348, 0.9813)$ is the 95% posterior interval for $R(200)$ whose Bayes estimate is $R^*(200) = 0.9614$.

Based on the censored sample, the MLE $\hat{R}(t) = \exp\left(-\frac{rt}{S_r}\right)$. At 200,

$\hat{R}(200) = \exp\left(\frac{-200}{5067.6}\right) = 0.9613 < R^*(200)$. Note that $\hat{R}(t) < R^*(t)$ for all

t since $\exp\left(-\dfrac{rt}{S_r}\right) > \left(1 + \dfrac{t}{S_r}\right)^{-r}$ for any $t > 0$.

We now consider similar problems for the two-parameter exponential distribution given by (8) with $\mu \geqslant 0$, which is a natural constraint in life-testing experiments.

Using a non-informative prior $g(\mu, \theta) \propto \dfrac{1}{\theta^a}$, $a > 0$, after some algebra we obtain the joint posterior of (μ, θ) as

$$\Pi(\mu, \theta \mid x) = \frac{k}{\theta^{r+a}} C_r \exp\left[-\{S + n(x_{(1)} - \mu)\}/\theta\right] \tag{12}$$

where

$$k = \frac{\Gamma(r + a - 2)}{n S^{r-a-2}}, \text{ and } C_r = \left(1 + \frac{nx_{(1)}}{S}\right)^{r+a-2} \Big/ \left[\left(1 + \frac{nx_{(1)}}{S}\right)^{r+a-2} - 1\right]$$

Note that here $S = \displaystyle\sum_{i=1}^{r} (x_{(i)} - x_{(1)}) + (n - r)(x_{(r)} - x_{(1)})$. Integrating out μ and θ from (12) we have the marginal posteriors of μ and θ as

$$\pi(\mu \mid x) = n(r + a - 2) C_r S^{-1}\left[1 + \frac{n(x_{(1)} - \mu)}{S}\right], \quad 0 < \mu < x_{(1)}$$

and

$$\pi(\theta \mid x) = \frac{C_r}{\theta^{r+a-1}} \frac{S^{r+a-2}}{\Gamma(r + a - 2)}\left[\exp\left(-\frac{S}{\theta}\right)\right]\left[1 - \exp\left(\frac{nx_{(1)}}{\theta}\right)\right], \quad 0 < \theta < \infty$$

Hence Bayes' estimators of μ and θ are

$$E(\mu \mid x) = \frac{C_r[n(r + a - 3)x_{(1)} - C_{r-1}^{-1}S]}{n(r + a - 3)}$$

$$E(\theta \mid x) = \frac{C_r S}{C_{r-1}(r + a - 3)}, \quad r + a > 3$$

Using the data in Grubbs (1971), $E(\mu \mid x)$ and $E(\theta \mid x)$ are tabulated in Table 2. For economy of space we do this only for one case, $r = n = 19$.

Table 6.2

	C_r			C_{r-1}			$E(\mu \mid x)$			$E(\theta \mid x)$		
a / r	0	1	2	0	1	2	0	1	2	0	1	2
19	1.052	1.043	1.036	1.057	1.047	1.039	118.47	120.02	121.57	987	930	879

We note that the expectations of μ are robust to changes in a, changing from 118.47 to 121.57, a difference of 2.6% while $E(\theta \mid x)$ has ratio of Min to Max of 0.89 (ratio is appropriate here since θ is a scale parameter).

We now turn to the question of estimating the reliability function

$$R_t = \exp\left(-\frac{t - \mu}{\theta}\right), \quad t > \mu$$

We assume that t's of interest for R_t are such that $t \geqslant x_{(1)}$. The posterior expectation of R_t is $E(R_t \mid x) = \int_0^{x_{(1)}} \int_0^{\infty} \exp\left(-\frac{t-\mu}{\theta}\right) \pi(\mu, \theta \mid x)d\theta \ d\mu =$ $\frac{n}{n+1} C_r \left[\left\{1 + \frac{t - x_{(1)}}{S}\right\}^{-(r+a-2)} - \left\{1 + \frac{t + nx_{(1)}}{S}\right\}^{-(r+a-2)}\right]$. Again we find that this Bayes' estimator of reliability to be quite robust in a and we illustrate the results in Table 6.3, compiled by using the data that lead to Table 6.2.

Table 6.3 $E(R_t \mid x)$ for the data of Table 6.1

		$r = 10$			$r = n = 19$	
t \ a	0	1	2	0	1	2
162	0.96	0.95	0.96	0.96	0.96	0.96
166	0.96	0.95	0.95	0.95	0.95	0.95
175	0.97	0.94	0.94	0.94	0.94	0.95
180	0.94	0.93	0.94	0.94	0.94	0.93
190	0.94	0.93	0.92	0.93	0.92	0.92

The posterior of R_t itself is often of interest. Use the transformation $\mu = t + \nu \log R_t$, $\theta = \nu$. The absolute value of the Jacobian of the transformation is ν/R_t. Substituting in (12) and integrating out ν we have the posterior of R_t given by

$$\pi(R_t \mid x) = C_r \frac{nS^{r+a-2n-1}}{\Gamma(r+a-2)} R_t \int_{\alpha_1(R_t)}^{\alpha_2(R_t)} \nu^{-(r+a-1)}$$
$$\exp\left[-\frac{1}{\nu}\{S + n(X_{(1)} - t)\}\right] d\nu$$

where

$$\alpha_1(R_t) = \frac{t - x_{(1)}}{\log \frac{1}{R_t}} \quad \text{and} \quad \alpha_2(R_t) = \frac{t}{\log \frac{1}{R_t}} \tag{13}$$

We define $\pi(0 \mid x) = \pi(1 \mid x) = 0$. This result for the case $a = 1$ coincides with a result of Pierce (1973) based on a different argument.

Using the identity connecting the incomplete gamma distribution with Poisson, we find that for the case $W_r = S + n(x_{(1)} - t) > 0$, $t > x_{(1)}$,

$$\pi(R_t \mid x) = \frac{nC_r S^{r+a-2}}{W_r^{r+a-2}} R_t^{n-1} \left[(R_t)\left(\frac{W_r}{t}\right)^{r+a-3} \sum_{m=0}^{} \left\{\frac{W_r \log \frac{1}{R_t}}{t}\right\}^m \frac{1}{m!}\right.$$
$$\left. -\{R_t\} \frac{W_r}{t - x_{(1)}}^{r+a-3} \sum_{m=0}^{} \left\{\frac{W_r \log \frac{1}{R_t}}{t - x_{(1)}}\right\}^m \frac{1}{m!}\right]$$

For the case $W_r < 0$, we refer to Sinha and Guttman (1976).

Using the aforementioned data of Grubbs, we plot in Figure 3 the posteriors of R_t for $t = 200$ and $r = n = 19$. We again note the robust-

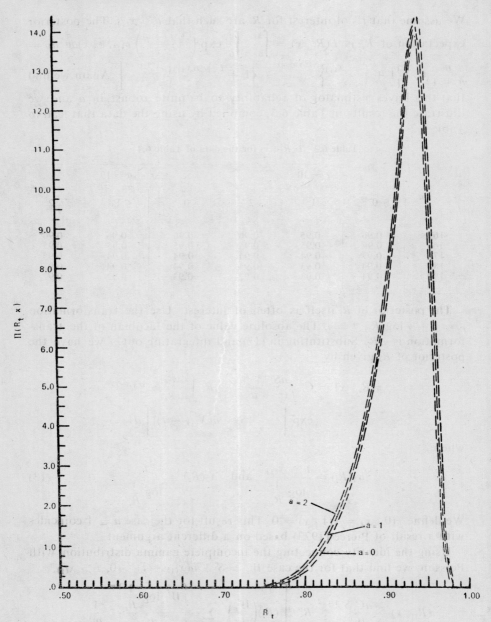

Fig. 3   Posterior of R_t at $t = 200$, $a = 1, 2, 3$

ness of the posteriors for changes in a.

To inquire further, we look at the $100(1 - \alpha)\%$ posterior limits (R_1, R_2) for R_t. Of course, it is now a problem of numerical integration to find R_1 and R_2 which are such that

$$P(R_t < R_1 \mid x) = P(R_t > R_2 \mid x) = \alpha/2 \text{ so that } P(R_1 < R_t < R_2) = 1 - \alpha$$

It is more convenient to use (13) directly for this purpose. Lower or upper one-sided posterior limits could be defined in the usual way. For Grubbs data, we have done the requisite numerical integration and produced lower posterior limits R_t^a which are such that $P(R_t^a \leqslant R_t \mid x) = = 1 - \alpha$ for $a = 0, 1, 2$. We have tabulated R_t^a for $1 - \alpha = 0.99, 0.95, 0.90$, $a = 0, 1, 2, r = n = 19$ in Table 6.4 for $t = 200$ and in Table 6.5 for $t = 180 (60) 600$. We again note the robustness of R_t^a for changes in a.

Table 6.4 Lower $100(1 - \alpha)\%$ **posterior limits of** R_t^a

a	$1 - \alpha$		
	0.99	0.95	0.90
0	0.8049	0.8443	0.8652
1	0.7987	0.8394*	0.8611
2	0.7928	0.8348	0.8572

*Pierce (1973) quotes 0.840 for this limit with the same set of data.

Table 6.5 Lower $100(1 - \alpha)\%$ **of posterior limits of** R_t^a

t	a	$1 - \alpha$		
		0.99	0.95	0.90
180	0	0.8263	0.8644	0.8851
	1	0.8208	0.8603	0.8819
	2	0.8156	0.8565	0.8789
240	0	0.7624	0.8039	0.8252
	1	0.7548	0.7974	0.8192
	2	0.7476	0.7911	0.8134
300	0	0.6994	0.7430	0.7647
	1	0.6900	0.7341	0.7562
	2	0.6809	0.7256	0.7479
360	0	0.6381	0.6831	0.7058
	1	0.6270	0.6723	0.6951
	2	0.6163	0.6617	0.6846
420	0	0.5795	0.6262	0.6502
	1	0.5670	0.6136	0.6376
	2	0.5549	0.6014	0.6254
480	0	0.5247	0.5730	0.5983
	1	0.5110	0.5591	0.5843
	2	0.4979	0.5456	0.5707
540	0	0.4740	0.5238	0.5503
	1	0.4596	0.5089	0.5351
	2	0.4457	0.4944	0.5204
600	0	0.4277	0.4785	0.5059
	1	0.4128	0.4628	0.4898
	2	0.3985	0.4477	0.4744

SOURCE: Sinha and Guttman (1976).

Weibull distribution

Consider the Weibull p.d.f. given by

$$f(x \mid p, \theta) = \frac{p}{\theta} x^{p-1} \exp(-x^p/\theta), \ x, p, \theta > 0 \tag{14}$$

Let the shape parameter p be known and let

$$g(\theta) \propto \frac{\exp(-a/\theta)}{\theta^c} \ a, c > 1, \quad \theta > 0 \tag{15}$$

be the prior distribution of θ. The posterior of θ

$$\Pi(\theta \mid x) = K \frac{\exp\left(-\frac{\Sigma x^p + a}{\theta}\right)}{\theta^{n+c}}$$

where

$$K^{-1} = \int_0^\infty \frac{\exp\left(-\frac{\Sigma x^p + a}{\theta}\right)}{\theta^{n+c}} \, d\theta = \frac{\Gamma(n+c-1)}{(\Sigma x^p + a)^{n+c-1}}$$

$$\Pi(\theta \mid x) = \frac{(\Sigma x^p + a)^{n+c-1}}{\Gamma(n+c-1)} \frac{\exp\left\{-\frac{\Sigma x^p + a}{\theta}\right\}}{\theta^{n+c}}, \theta > 0 \tag{16}$$

and Bayes estimator of θ is

$$\theta^* = \frac{\Sigma x^p + a}{n + c - 2}$$

which for large sample (in relation to a and c) will be quite close to the MLE $\hat{\theta} = \frac{\Sigma x^p}{n}$. For $a = 0$, $c = 2$ in (15), the Bayes estimator of θ is the same as the MLE of θ.

EXAMPLE 6.3 A sample of 25 observations was generated from the Weibull distribution (2.1) with $p = 2$, $\theta = 4$.

SAMPLE: x

1.8487	1.8802	1.7961	1.6560	1.3162
0.3761	1.5700	0.5319	1.7172	0.7705
0.7500	1.7708	0.5903	1.9310	1.8889
3.0530	1.3592	0.6288	1.0509	4.1505
1.3545	3.0466	0.6461	1.6173	1.8889

For the above sample, Table 6.6 gives θ^* for $a = 0$, (2), 8 and $c = (0, 1)$, 3.

Table 6.6 (p known) θ^*: **Bayes estimates of** θ

c	a					
	0	2	4	6	8	$\left(\dfrac{\text{Min}}{\text{Max}}\middle/ c\right)$
0	3.5698	3.6567	3.7437	3.8306	3.9176	0.9112
1	3.4210	3.5044	3.5877	3.6710	3.7544	0.9112
2	3.2842	3.3642	3.4442	3.5242	3.6042	0.9112
3	3.1579	3.2348	3.3117	3.3886	3.4656	0.9112
$\left(\dfrac{\text{Min}}{\text{Max}}\middle/ a\right)$	0.8846	0.8846	0.8846	0.8846	0.8846	

The posterior distributions (16) are plotted in Figures 4 to 7. The posteriors and the corresponding expectations (Table 6.6) appear to be quite robust for changes in a.

Consider now the case when p is also unknown. If we use Jeffrey's invariant prior, this leads to quite intractable calculations. Note that $\theta^{1/p}$ is the scale parameter and p is the shape parameter and it is customary to assume that the prior distribution of (θ, p) is $g(\theta)h(p)$ where $g(\theta) \propto (1/\theta^c)$, $c > 0$ and $h(p) \propto (1/a)$, $0 < p < a$. With these priors we obtain the posterior distribution of (θ, p).

$$\Pi(\theta, p \mid \mathbf{x}) = \frac{kp^n}{\theta^{n+c}} \lambda^{p-1} \exp\left(\frac{\sum_1^n x_i^p}{\theta}\right) \tag{17}$$

where

$$\lambda = \Pi_{i=1}^n x_i$$

and

$$k^{-1} = \int_0^a \int_0^\infty \Pi(\theta, p \mid \mathbf{x}) \, d\theta \, dp = \Gamma(n + c - 1) \int_0^a \frac{p^n \lambda^{p-1}}{\left(\sum_1^n x_i^p\right)^{n+c-1}} \, dp$$

Integrating out θ in (17) we have the marginal posterior of p:

$$\Pi(p \mid \mathbf{x}) = \frac{\dfrac{p^n \lambda^{p-1}}{\left(\sum_1^n x_i^p\right)^{n+c-1}}}{\displaystyle\int_0^a \frac{p^n \lambda^{p-1}}{(\sum x_i^p)^{n+c-1}} \, dp}. \tag{18}$$

We expect a more robust pattern in the posteriors of p for small values of a. The posteriors (18) plotted in Figures 8 to 11 for $a = 4(2)10$ with $c = 0(1)3$ support such a pattern.

Similarly we obtain the marginal posterior of θ, viz.,

$$\Pi(\theta \mid x) = \frac{\frac{1}{\theta^{n+c}} \int_0^a p^n \lambda^{p-1} \exp\left(-\sum_1^n x_i^p / \theta\right) dp}{\Gamma(n+c-1) \int_0^a \frac{p^n \lambda^{p-1}}{\left(\sum_1^n x_i^p\right)^{n+c-1}} dp} \tag{19}$$

We expect erratic behaviour of these posteriors for larger values of a and the posteriors in Figures 12 to 14 indicate such a pattern.

From (18) and (19) we obtain the corresponding Bayes estimators

$$p^* = \frac{\int_0^a p^{n+1} \lambda^{p-1} \Big/ \left(\sum_1^n x_i^p\right)^{n+c-1} dp}{\int_0^a p^n \lambda^{p-1} \Big/ \left(\sum_1^n x_i^p\right)^{n+c-1} dp}$$

and

$$\theta^* = \frac{1}{n+c-2} \frac{\int_0^a p^n \lambda^{p-1} \Big/ \left(\sum_1^n x_i^p\right)^{n+c-2} dp}{\int_0^a p^n \lambda^{p-1} \Big/ \left(\sum_1^n x_i^p\right)^{n+c-1} dp}$$

Table 6.7 p^*: Bayes estimates of p

c	a				
	4	6	8	10	$\left(\frac{\text{Min}}{\text{Max}} \Big/ c\right)$
0	2.1386	2.1135	2.2008	2.0236	0.9195
1	2.0606	2.0273	2.0953	1.9647	0.9377
2	1.9893	1.9593	1.9853	1.9258	0.9681
3	1.9239	1.9058	1.8804	1.8995	0.9772
$\left(\frac{\text{Min}}{\text{Max}} \Big/ a\right)$	0.8996	0.9017	0.8544	0.9387	

Table 6.8 θ^*: Bayes estimates of θ

c	a				
	4	6	8	10	$\left(\frac{\text{Min}}{\text{Max}} \Big/ c\right)$
0	4.1929	4.1416	4.4516	3.8015	0.8539
1	3.7473	3.6571	3.9214	3.4219	0.8726
2	3.3794	3.2887	3.4404	3.1517	0.9132
3	3.0174	3.0065	3.0275	2.9492	0.9680
$\left(\frac{\text{Min}}{\text{Max}} \Big/ a\right)$	0.7325	0.7259	0.6801	0.7758	

A comparison of Tables 6.7 and 6.8 shows that p^* and θ^* behave much the same way. The posteriors of θ are far from robust for changes in a.

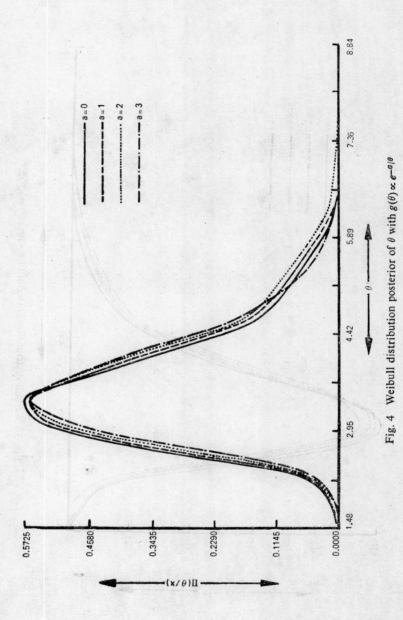

Fig. 4   Weibull distribution posterior of θ with $g(\theta) \propto e^{-a/\theta}$

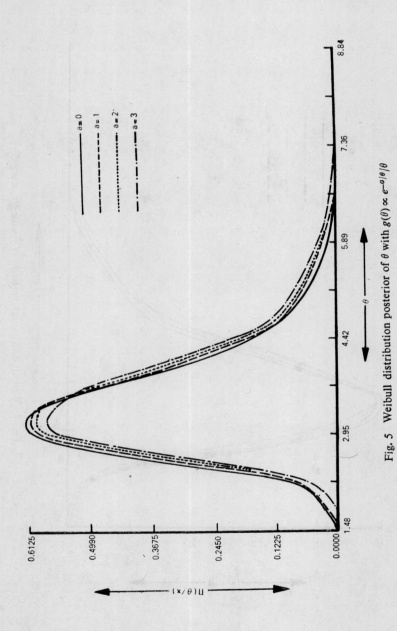

Fig. 5   Weibull distribution posterior of θ with $g(\theta) \propto e^{-a/\theta}/\theta$

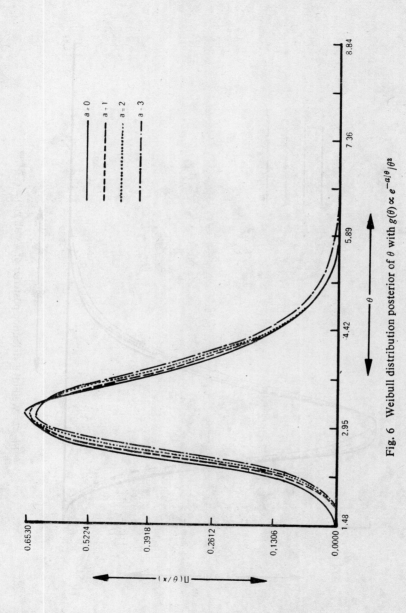

Fig. 6 Weibull distribution posterior of θ with $g(\theta) \propto e^{-a/\theta}/\theta^2$

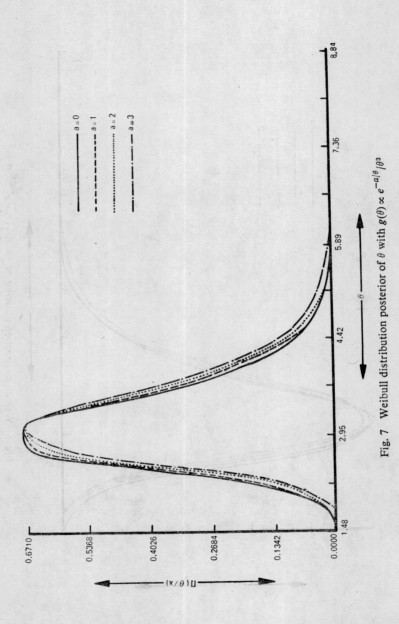

Fig. 7 Weibull distribution posterior of θ with $g(\theta) \propto e^{-a/\theta}/\theta^3$

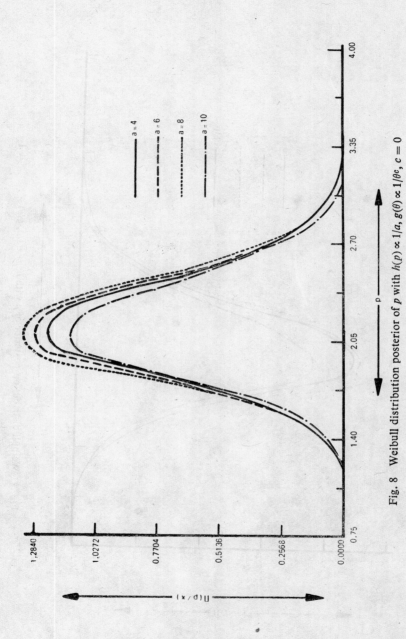

Fig. 8 Weibull distribution posterior of p with $h(p) \propto 1/a$, $g(\theta) \propto 1/\theta^c$, $c = 0$

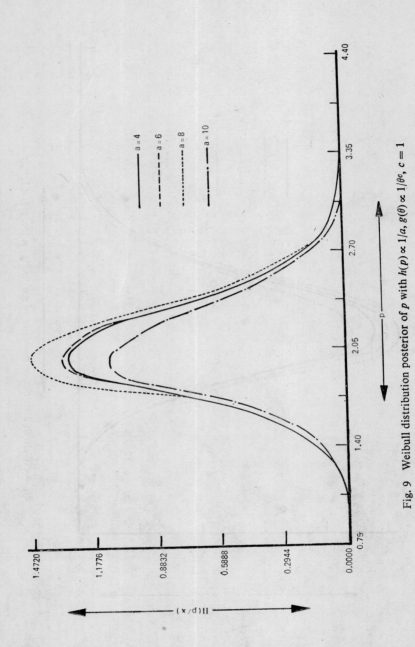

Fig. 9   Weibull distribution posterior of p with $h(p) \propto 1/a$, $g(\theta) \propto 1/\theta c$, $c = 1$

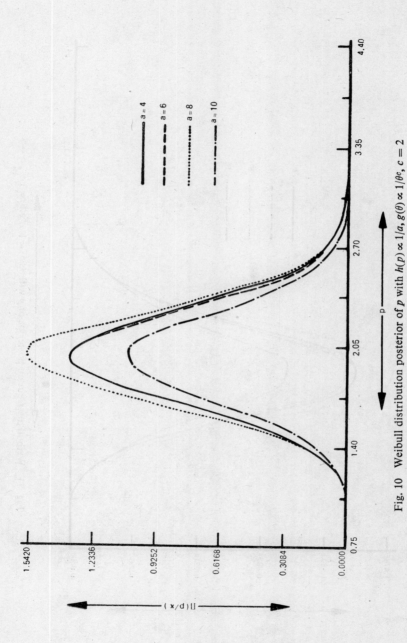

Fig. 10 Weibull distribution posterior of p with $h(p) \propto 1/a$, $g(\theta) \propto 1/\theta^c$, $c = 2$

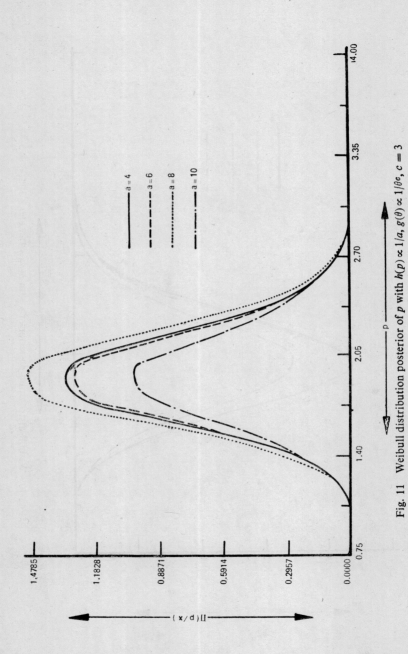

Fig. 11   Weibull distribution posterior of p with $h(p) \propto 1/a$, $g(\theta) \propto 1/\theta c$, $c = 3$

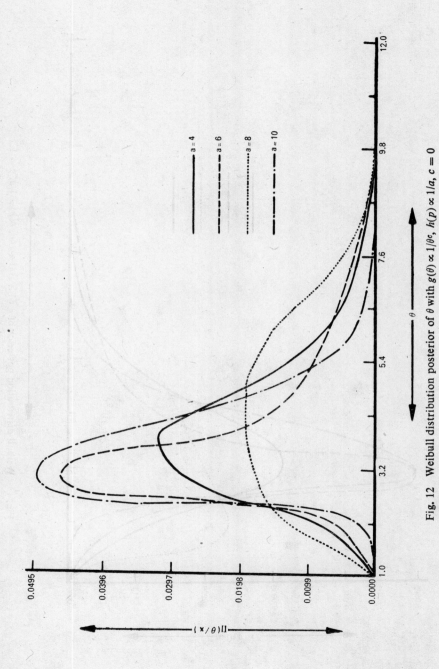

Fig. 12 Weibull distribution posterior of θ with $g(\theta) \propto 1/\theta c$, $h(p) \propto 1/a$, $c = 0$

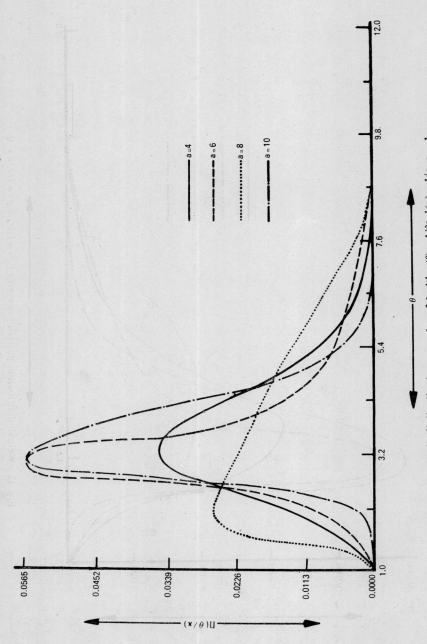

Fig. 13   Weibull distribution posterior of θ with $g(\theta) \propto 1/\theta^c$, $h(p) \propto 1/a$, $c = 1$

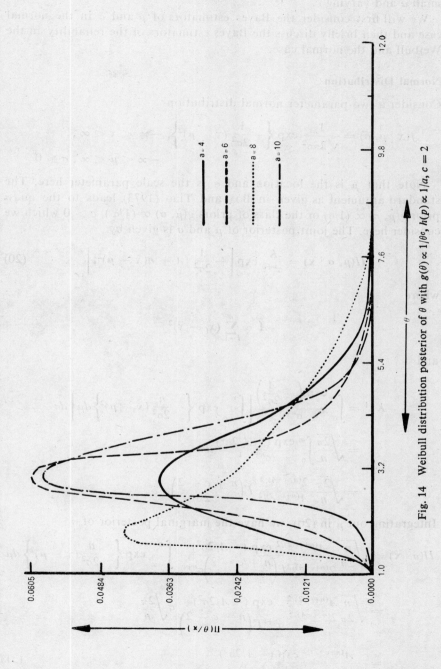

Fig. 14 Weibull distribution posterior of θ with $g(\theta) \propto 1/\theta c$, $h(p) \propto 1/a$, $c = 2$

Thus θ^* is much more non-robust than p^* while both are fairly robust for small a and varying c.

We will first consider the Bayes estimators of μ and σ in the normal case and then briefly discuss the Bayes estimators of the reliability in the Weibull and the normal case.

Normal Distribution

Consider a two-parameter normal distribution

$$f(x \mid \mu, \sigma) = \frac{1}{\sqrt{2\pi\sigma^2}} \exp\left\{-\frac{1}{2\sigma^2}(x-\mu)^2\right\}, \quad -\infty < x < \infty,$$

$$-\infty < \mu < \infty, \sigma > 0$$

Note that μ is the location and σ is the scale parameter here. The standard argument as given in Box and Tiao (1973) leads to the quasi prior $g(\mu, \sigma) \propto (1/\sigma)$ or the class of priors $g(\mu, \sigma) \propto (1/\sigma^c)$, $c > 0$ which we consider here. The joint posterior of μ and σ is given by

$$\Pi(\mu, \sigma \mid \mathbf{x}) = \frac{K}{\sigma^{n+c}} \exp\left[-\frac{1}{2\sigma^2}\{A + n(\bar{x} - \mu)^2\}\right] \tag{20}$$

where

$$A = \sum_{i=1}^{n} (x_i - \bar{x})^2$$

and

$$K^{-1} = \int_0^\infty \frac{\exp\left(-\dfrac{A}{2\sigma^2}\right)}{\sigma^{n+c}} \left[\int_{-\infty}^{\infty} \exp\left\{-\frac{n}{2\sigma^2}(\bar{x} - \mu)^2\right\} d\mu\right] d\sigma$$

$$= \sqrt{\frac{2\pi}{n}} \int_0^\infty \frac{\exp(-A/2\sigma^2)}{\sigma^{n+c-1}} d\sigma$$

$$= \sqrt{\frac{2\pi}{n}} \cdot \frac{2^{(n+c-4)/2}}{A^{(n+c-2)/2}} \Gamma\left(\frac{n+c-2}{2}\right).$$

Integrating out μ in (20), we have the marginal posterior of σ

$$\Pi(\sigma \mid \mathbf{x}) = \sqrt{\frac{n}{2\pi}} \frac{A^{(n+c-2)/2} \exp(-A/2\sigma^2)}{2^{(n+c-4)/2}\Gamma\left(\dfrac{n+c-2}{2}\right)\sigma^{n+c}} \int_{-\infty}^{\infty} \exp\left\{-\frac{n}{2\sigma^2}(\bar{x} - \mu)^2\right\} d\mu$$

$$= \sqrt{\frac{n}{2\pi}} \frac{A^{(n+c-2)/2}}{2^{(n+c-4)/2}} \frac{\exp(-A/2\sigma^2)}{\sigma^{n+c}\Gamma\left(\dfrac{n+c-2}{2}\right)} \sigma\sqrt{\frac{2\pi}{n}}$$

$$= \frac{A^{(n+c-2)/2} \exp(-A/2\sigma^2)}{2^{(n+c-4)/2}\Gamma\left(\dfrac{n+c-2}{2}\right)\sigma^{n+c-1}} \tag{21}$$

Similarly,

$\Pi(\mu \mid x) =$ the marginal posterior of μ

$$= \sqrt{\frac{n}{2\pi}} \left(\frac{A}{2}\right)^{(n+c-2)/2} \cdot \frac{1}{\Gamma\left(\dfrac{n+c-2}{2}\right)} \int_0^\infty \frac{\exp\left[-\dfrac{1}{2\sigma^2}\{A + n(\bar{x}-\mu)^2\}\right]}{(\sigma^2)^{\frac{1}{2}(n+c-1)+1}} \, d\sigma^2$$

$$= \sqrt{\frac{n}{2\pi}} \left(\frac{A}{2}\right)^{(n+c-2)/2} \cdot \frac{1}{\Gamma\left(\dfrac{n+c-2}{2}\right)} \frac{2^{(n+c-1)/2}\Gamma\left(\dfrac{n+c-1}{2}\right)}{[A + n(\bar{x}-\mu)^2]^{(n+c-1)/2}}$$

$$= \sqrt{\frac{n}{A}} \frac{\Gamma\left(\dfrac{n+c-1}{2}\right)}{\sqrt{\pi}\,\Gamma\left(\dfrac{n+c-2}{2}\right)} \frac{1}{\left[1 + \dfrac{n(\bar{x}-\mu)^2}{A}\right]^{(n+c-1)/2}}$$

$$= \sqrt{\frac{n}{A}} \frac{1}{B\left(\dfrac{1}{2}, \dfrac{n+c-2}{2}\right)} \frac{1}{\left[1 + \dfrac{n(\bar{x}-\mu)^2}{A}\right]^{(n+c-1)/2}}, \qquad -\infty < \mu < \infty \qquad (22)$$

From (21) we obtain the Bayes estimator of σ,

$$\sigma^* = E(\sigma \mid x)$$

$$= \left(\frac{A}{2}\right)^{(n+c-2)/2} \frac{1}{\Gamma\left(\dfrac{n+c-2}{2}\right)} \int_0^\infty \frac{\exp\left(-\dfrac{A}{2\sigma^2}\right) d\sigma^2}{(\sigma^2)^{\frac{1}{2}(n+c-3)+1}}$$

$$= \sqrt{\frac{A}{2}} \frac{\Gamma\left(\dfrac{n+c-3}{2}\right)}{\Gamma\left(\dfrac{n+c-2}{2}\right)}$$

Note that Bayes estimator of σ^2, viz.,

$$\sigma^{*2} = \int_0^\infty \sigma^2 \Pi(\sigma \mid x) \, d\sigma$$

$$= \frac{A}{n+c-4}$$

We note that the MLE of σ^2 coincides with the σ^{*2} for $c = 4$ and the UMVUE of σ^2 is the same as Bayes estimator for σ^2 when $c = 3$. As in the exponential case $\sigma^{*2}(c_1)$ and $\sigma^{*2}(c_2)$ differ from each other only by a small number if n is large. Indeed, $|\sigma^{*2}(c_1) - \sigma^{*2}(c_2)| \to 0$ in probability as $n \to \infty$. From (22) we have

$$\mu^* = \sqrt{\frac{n}{A}} \frac{1}{B\left(\dfrac{1}{2}, \dfrac{n+c-2}{2}\right)} \int_{-\infty}^\infty \frac{\mu \, d\mu}{\left[1 + \dfrac{n(\bar{x}-\mu)^2}{A}\right]^{(n+c-1)/2}}$$

$$= \sqrt{\frac{n}{A}} \frac{1}{B\left(\frac{1}{2}, \frac{n+c-2}{2}\right)} \int_{-\infty}^{\infty} \frac{\left(\bar{x} - \frac{s}{\sqrt{n}} t\right) \frac{s}{\sqrt{n}} dt}{\left(1 + \frac{t^2}{n-1}\right)^{(n+c-1)/2}} = \bar{x}$$

Thus μ^* is same as the MLE and UMVUE of μ for any choice of $c > 0$.

EXAMPLE 6.4   A random sample of 30 observations was generated from a normal population with $\mu = 20$ and $\sigma = 5$.

SAMPLE : x

22.320	24.530	17.590	11.065	10.975
20.300	25.895	13.120	19.475	14.070
27.430	12.495	14.950	13.305	23.290
25.110	16.550	19.975	25.205	17.805
26.970	26.860	26.965	21.395	13.005
20.995	20.795	31.365	20.205	14.340

We have $\bar{x} = 19.945$, $A = 931.0165$, $n = 30$.

The posteriors (21) and (22) were plotted for $c = 0, 1, 2, 3$ for a wide range of values of μ and σ. Figures 15 and 16 show that the posteriors of μ are very much robust for variations in c, while those of σ are less so. Similar property of robustness is reflected in the corresponding Bayes estimators of μ and σ (Sinha, 1979).

Table 6.9   Bayes Estimates of σ

c	σ^*	$\left(\dfrac{\text{Min}}{\text{Max}}\right)$
0	5.9268	
1	5.8180	0.9478
2	5.7151	
3	5.6174	

Reliability Estimation

We have discussed earlier Bayes estimate of reliability for one and two-parameter exponential distributions. We will briefly consider the Weibull and the normal case. For the Weibull p.d.f. (14), Bayes estimator of reliability is given by

$$R_t^* = E\left\{\exp\left(-\frac{t^p}{\theta}\right) \middle| \mathbf{x}\right\}$$

$$= \iint \pi(\theta, p \mid \mathbf{x}) \exp\left(\frac{-t^p}{\theta}\right) d\theta \, dp$$

If p is known, under the prior and posterior distributions of θ as specified

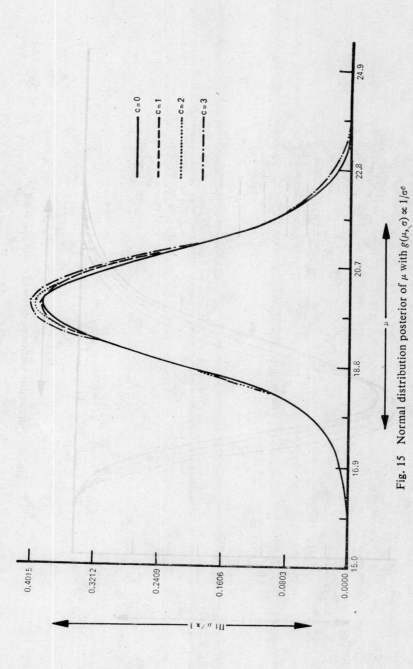

Fig. 15 Normal distribution posterior of μ with $g(\mu, \sigma) \propto 1/\sigma^c$

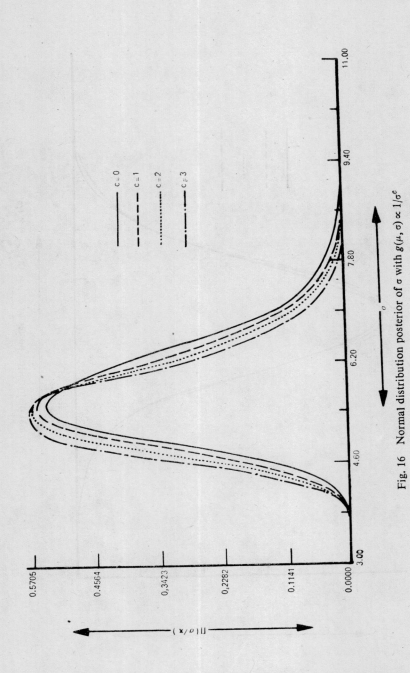

Fig. 16   Normal distribution posterior of σ with $g(\mu, \sigma) \propto 1/\sigma^c$

in (15) and (16), we obtain

$$R_t^* = \frac{1}{\left(1 + \dfrac{t^p}{\Sigma x^p + a}\right)^{n+c-1}} \tag{23}$$

When p is unknown, we use the joint posterior of θ and p assumed earlier, namely $g(\theta)h(p) \propto \dfrac{1}{\theta^c}\dfrac{1}{a}$. After some algebra we obtain from (17)

$$R_t^* = \frac{\displaystyle\int_0^a \frac{p^n \lambda^{p-1}}{(\Sigma x^p + t^p)^{n+c-1}}\, dp}{\displaystyle\int_0^a \frac{p^n \lambda^{p-1}}{(\Sigma x^p)^{n+c-1}}\, dp} \tag{24}$$

For the data in example 6.3, we compute (23) and (24) at $t = 2$ for different values of c and a.

Table 6.10   ($p = 2$)

c	a		
	2	4	6
0	0.3190	0.3276	0.3360
1	0.3042	0.3127	0.3211
2	0.2900	0.2985	0.3068
3	0.2766	0.2849	0.2932

Table 6.11   (p unknown)

c	a		
	2	4	6
0	0.3028	0.3124	0.3030
1	0.2876	0.2964	0.2875
2	0.2781	0.2816	0.2733
3	0.2592	0.2676	0.2599

For the normal distribution $N(\mu, \sigma^2)$, the reliability at time t is given by

$$
\begin{aligned}
R_t &= 1 - \frac{1}{\sqrt{2\pi}\sigma}\int_t^\infty \exp\left(\frac{-(x-\mu)^2}{2\sigma^2}\right) dx \\
&= 1 - \frac{1}{\sqrt{2\pi}}\int_{(t-\mu)/\sigma}^\infty \exp\left(\frac{-\mu^2}{2}\right) du \\
&= 1 - \Phi\left(\frac{t-\mu}{\sigma}\right)
\end{aligned}
$$

Thus, Bayes estimator

$$R_t^* = 1 - \int_{-\infty}^\infty \int_0^\infty \Phi\left(\frac{t-\mu}{\sigma}\right)\pi(\mu, \sigma \mid \mathbf{x})\, d\sigma\, d\mu$$

where $\pi(\mu, \sigma \mid \mathbf{x})$ is given by (20).

EXERCISES

6.1 $x = (x_1, x_2, \ldots, x_n)$ is a random sample from the population with p.d.f. $f(x, \lambda) = \lambda \exp(-x\lambda)$, $x > 0$, $\lambda > 0$. Compute Bayes estimators of λ under the following priors of λ:

(i) $g(\lambda) = \mu^\nu \exp(-\lambda\mu) \dfrac{\lambda^{\nu-1}}{\Gamma(\nu)}$, $\nu > 0$

(ii) $g(\lambda) = \dfrac{1}{\lambda}$

6.2 15 electric bulbs under test failed at the following hours:

22.67, 25.35, 33.72, 39.18, 44.12, 49.71, 58.65, 67.20, 70.10, 81.07, 107.58, 113.73, 120.04, 142.86, 148.99.

If the life distribution is given by

$$f(x, \theta) = \frac{1}{\theta} \exp\left(-\frac{x}{\theta}\right), \quad x, \theta > 0$$

and θ is treated as a random variable with a p.d.f.

$$g(\theta) = \frac{1}{\theta^2} \exp\left(-\frac{1}{\theta}\right), \quad \theta > 0.$$

Compute Bayes estimators of θ and reliability function at $t = 150$ hours.

6.3 Consider the following failure times arranged as order statistics

15.68, 18.71, 24.14, 26.33, 35.72, 49.18, 54.36, 68.27, 77.20, 91.07, 123.59, 152.68, 159.00, 162.15, 193.62.

Let the distribution be given by a two-parameter exponential p.d.f.

$$f(x \mid \theta, \mu) = \frac{1}{\theta} \exp\left\{-\left(\frac{x - \mu}{\theta}\right)\right\}, \quad x \geqslant \mu, \theta > 0.$$

Obtain Bayes estimates of R_t at $t = 100$ hours under the following joint priors:

(i) $p(\mu, \theta) = \dfrac{1}{\theta}$, $\theta > 0$, $-\infty < \mu < \infty$

(ii) $p(\mu, \theta) \propto \theta$.

and compare your estimates with the corresponding UMVUE of R_t.

6.4 Let the random variable $X \sim N(\theta, \sigma^2)$ where prior (θ, σ^2),

$$p(\theta, \sigma^2) = \frac{1}{\sqrt{2\pi k^2}} \exp\left[-\frac{(\theta - \mu)^2}{2k^2}\right]$$

(a) Obtain Bayes estimator of (θ, σ).

(b) Compute $R_t = P(X > t)$ in terms of the cumulative distribution function

$$\Phi(u) = \frac{1}{\sqrt{2\pi}} \int_{-\infty}^{u} \exp\left(-\frac{z^2}{2}\right) dz$$

6.5 20 items under test failed at the following hours:

10.99, 15.79, 24.14, 34.43, 43.72, 51.72, 56.12, 68.27, 77.20, 88.47, 91.07, 117.58, 130.40. 133.12, 152.90, 159.00, 193.62, 208.71, 308.82, 316.07.

Let the failure time distribution be given by:

$$f(x \mid \mu, \theta) = \frac{1}{\theta} \exp\left\{-\left(\frac{x - \mu}{\theta}\right)\right\}, \quad x \geqslant \mu, \theta > 0$$

Compute Bayes estimates of (μ, θ) and R_t at $t = 100$ hours under the joint prior $p(\mu, \theta) = 1$, based on

(a) all the 20 observations,

(b) the first 10 observations.

CHAPTER 7

Reliability of Series/Parallel Systems

Consider a system which consists of k different components $c_1, c_2, \ldots, c_k$ connected in series. The system fails as soon as any one of the components fails. The reliability of the system $R_t(s, k)$ is the probability that none of the components fails before time t, i.e.,

$$R_t(s, k) = P\{c_1 \geqslant t, c_2 \geqslant t, \ldots, c_k \geqslant t\}$$

On the assumption that the failure time distributions of the components are independent,

$$
\begin{aligned}
R_t(s, k) &= P(c_1 \geqslant t) \cdot P(c_2 \geqslant t) \ldots P(c_k \geqslant t) \\
&= R_1(t) \cdot R_2(t) \ldots R_k(t) \\
&= \prod_{i=1}^{k} R_i(t)
\end{aligned}
$$

where $R_i(t)$ denotes the reliability of the component c_i $(i = 1, 2, \ldots, k)$. If the failure times of the components are identically distributed,

$$R_t(s, k) = [R(t)]^k.$$

We note that since $R_i(t) \leqslant 1$, we have $R_t(s, k) \leqslant \underset{1 \leqslant i \leqslant k}{\text{Min}} R_i(t)$ and the reliability of a system in which the components are connected in series is much smaller than the reliability of the individual units.

Similarly, one can consider a system consisting of components $c_1, c_2, \ldots, c_k$ arranged not in series but are parallel. Thus, the system will not fail if even one of the components is working, or equivalently the system fails when each of the component fails. Let $R_t(p, k)$ denote the corresponding reliability for a parallel system. Then

$$
\begin{aligned}
1 - R_t(p, k) &= \text{Prob (system fails before } t) \\
&= P(c_1 \leqslant t, c_2 \leqslant t, \ldots, c_k \leqslant t).
\end{aligned}
$$

If the failure time distributions are independent,

$$1 - R_t(p, k) = \prod_{i=1}^{k} F_i(t) = \prod_{i=1}^{k} [1 - R_i(t)]$$

or

$$R_t(p, k) = 1 - \prod_{i=1}^{k} [1 - R_i(t)]$$

We can show that $R_t(p, k) \geqslant \underset{1 \leqslant i \leqslant k}{\text{Max}} R_i(t)$ by noting that

$$R_t(p, k) = P(A_1 \cup A_2 \ldots \cup A_k) \geqslant \underset{1 \leqslant i \leqslant k}{\text{Max}} P(A_i)$$

where A_i is the event that the ith component does not fail before time t.

In this chapter we will study the reliability problems for a series and parallel system with k independent components. In view of the importance of the exponential distribution, we will assume that the failure time distributions have one- or two-parameter exponential densities.

Series System with Identical Components

Let us consider a system arranged in series and such that each c_i has failure time distribution given by one-parameter exponential density with p.d.f.

$$f(x, \sigma) = \frac{1}{\sigma} \exp\left(-\frac{x}{\sigma}\right), \quad x \geqslant 0, \sigma > 0$$

Then,

$$R_t(s, k) = [R_t(\sigma)]^k = \exp\left\{-\frac{kt}{\sigma}\right\}$$

and if S denotes the failure time of the system then

$$P\{S \leqslant t\} = 1 - \exp\left\{-\frac{kt}{\sigma}\right\}, \quad t > 0, \sigma > 0$$

This shows that the failure time distribution of the system S is itself exponential with p.d.f.

$$f(y, \theta) = \frac{1}{\theta} \exp\left\{-\frac{y}{\theta}\right\}, \quad y > 0, \theta = \frac{\sigma}{k} > 0$$

If n such systems are put on test with observed failure times as $(y_1, y_2, \ldots, y_n)$, then the results from Chapter 1 immediately give

$$\hat{\theta} = \frac{1}{n} \sum_{i=1}^{n} y_i = \bar{y}$$

and the MLE of the $R_t(s, k) = R_t(\theta) = \exp\{-t/\theta\}$ is given by

$$\hat{R}_t(s, k) = \exp\left\{-\frac{t}{\theta}\right\} = \exp\left\{-\frac{nt}{S_n}\right\}$$

where $S_n = \sum_{i=1}^{n} y_i$. Similarly, the UMVUE of $R_t(s, k)$ is given by

$$\tilde{R}_t(s, k) = \left(1 - \frac{t}{S_n}\right)^{n-1} \quad \text{if } S_n > t$$

$$= 0 \qquad\qquad\qquad \text{if } S_n \leqslant t$$

The results for the failure censored sample can be obtained by replacing n by r and S_n by $S_r = \sum_{i=1}^{r} y_{(i)} + (n - r)y_{(r)}$.

Next, consider the case where $(c_1, c_2, \ldots, c_k)$ are i.i.d. r.v. with each c_i having failure time distribution which is two-parameter exponential. Thus, the p.d.f. of each c_i is given by

$$f(x, \mu, \sigma) = \frac{1}{\sigma} \exp\left\{-\frac{x-\mu}{\sigma}\right\}, \quad x \geqslant \mu, \sigma > 0$$

Here

$$R_t(s, k) = \exp\left\{-\frac{k(t-\mu)}{\sigma}\right\}, \quad t \geqslant \mu$$

$$= 1 \qquad t < \mu$$

It immediately follows that

$$P\{S \leqslant t\} = 1 - \exp\left\{-\frac{k(t-\mu)}{\sigma}\right\}, \quad t \geqslant \mu$$

$$= 0 \qquad t < \mu$$

The failure time distribution of S is itself two-parameter exponential with μ and $\theta = \sigma/k$, i.e., with p.d.f.

$$f(y, \mu, \theta) = \frac{1}{\theta} \exp\left\{-\frac{(y-\mu)}{\theta}\right\}, \quad y \geqslant \mu, \theta = \frac{\sigma}{k} > 0$$

If we have n systems $s_1, s_2, \ldots, s_n$ put on test with observed failure times $y_1, y_2, \ldots, y_n$, then the results of Chapter 1 on the two-parameter exponential model immediately yield $\hat{\mu} = y_{(1)}$ and $\hat{\theta} = \bar{y} - y_{(1)}$ as the MLE's of μ and $\theta = \sigma/k$. The MLE of $R_t(s, k) = R_t(\mu, \theta)$ is given by

$$R_t(\hat{\mu}, \hat{\theta}) = \exp\left\{-\frac{(t-y_{(1)})}{(\bar{y}-y_{(1)})}\right\} \quad \text{if } t > y_{(1)}$$

$$= 1 \qquad \text{if } t \leqslant y_{(1)}$$

As in Chapter 1, we will not go into the detailed derivation of UMVUE of $R_t(s, k) = R_t(\mu, \theta)$ but mention that following the methods and results given by Laurent (1963), we can derive the UMVUE of $R_t(\mu, \theta)$ as

$$\tilde{R}_t(\mu, \theta) = 1 \qquad \text{if } t < y_{(1)}$$

$$= \left(1 - \frac{1}{n}\right)\left[1 - \frac{t-y_{(1)}}{v_n}\right]^{n-2}, \quad y_{(1)} \leqslant t < v_n + y_{(1)}$$

$$= 0 \qquad \text{if } t \geqslant v_n + y_{(1)}$$

where

$$v_n = \sum_{i=2}^{n} [y_{(i)} - y_{(1)}]$$

If we have a failure censored sample the results can be obtained by replacing n by r and v_n by v_r where

$$v_r = \sum_{i=1}^{r} [y_{(i)} - y_{(1)}] + (n-r)(y_{(r)} - y_{(1)})$$

We note that if n systems are put on test, we are using $N = nk$ components and this may be an expensive requirement. We may view the problem in a slightly different way and obtain estimators for $R_t(s, k)$ with only m number of components put on test, rather than putting n systems on test. Thus, we now assume that m components are put on test with observed failure times $(x_1, x_2, \ldots, x_m)$ and the failure time distribution is one-parameter exponential with mean life σ. Thus, $(x_1, x_2, \ldots, x_m)$ is a random sample from a population with p.d.f. $(1/\sigma) \exp(-x/\sigma)$, $x \geqslant 0$, $\sigma > 0$, and the parametric function that we want to estimate is $R_t(s, k) = \exp\{-(kt/\sigma)\}$. Now we can use all the results of Chapter 1 by noting that we are estimating the reliability $R_t(\sigma) = \exp\{-t'/\sigma\}$ where $t' = kt$. Thus the MLE and the UMVUE of $R_t(s, k)$ are now given by

$$\hat{R}_t(s, k) = \exp\left\{-\frac{kt}{\bar{x}}\right\} = \exp\left\{-\frac{mkt}{S_m}\right\}$$

and

$$\tilde{R}_t(s, k) = \left(1 - \frac{kt}{S_m}\right)^{m-1} \quad \text{if } S_m \geqslant kt$$

$$= 0 \qquad\qquad \text{otherwise}$$

where $S_m = \sum_{i=1}^{m} x_i$. Again, if we are considering the failure censored sample, we can obtain the corresponding estimators by substituting r for m and S_r for S_m where

$$S_r = \sum_{i=1}^{r} x_{(i)} + (m - r)x_{(r)}$$

Reliability Bounds—Classical Approach

Suppose we have a system S with k components connected in series. The components are independently but not identically distributed, i.e., each c_i is distributed as a single-parameter exponential with mean life θ_i.

$$R_t(s, k) = \prod_{i=1}^{k} R_i(t) = \exp\left\{-t\left(\frac{1}{\theta_1} + \frac{1}{\theta_2} + \cdots + \frac{1}{\theta_k}\right)\right\}$$

Suppose we put n_i items of the type c_i to test and want to make inferences about $\phi = \sum_{i=1}^{k} \frac{1}{\theta_i}$. This problem is quite complicated for general k and we will study the case for $k = 2$.

Let $(x_1, x_2, \ldots, x_n)$ and $(y_1, y_2, \ldots, y_m)$ be the failure times of the components of the types c_1 and c_2 respectively and let the failure time distributions of c_i be independent with p.d.f.s

$$f(x, \theta_i) = \frac{1}{\theta_i} \exp\left(-\frac{x}{\theta_i}\right), \quad x, \theta_i > 0, \, i = 1, 2, \ldots \tag{1}$$

The parameter $\phi = (1/\theta_1) + (1/\theta_2)$ is of interest in that

$$R_t(s, 2) = \exp(-t\phi)$$

Let

$$z_1 = \sum_{i=1}^{n} x_i, \quad z_2 = \sum_{i=1}^{m} y_i$$

Then z_1/θ_1 and z_2/θ_2 are independent gamma variates with joint p.d.f.

$$g(z_1, z_2, \theta_1, \theta_2) = K \exp\left(-\frac{z_1}{\theta_1} - \frac{z_2}{\theta_2}\right) z_1^{n-1} z_2^{m-1} \qquad (2)$$

where

$$K = \frac{1}{\Gamma(n)\Gamma(m)\theta_1^{n-1}\theta_2^{m-1}}$$

Lentner and Buehler (1963) showed that the conditional distribution of z_1, given $z_1 - z_2 = u$, can be used to make inferences about ϕ.

Let $z_1 = w$, $z_1 - z_2 = u$, so that $z_1 = w$, $z_2 = w - u$.

$$h(w, u, \phi) = K \exp\left(-w\phi + \frac{u}{\theta_2}\right) w^{n-1}(w - u)^{m-1}, \quad w > 0, \; w > u$$

The marginal density of u is given by

$$h_1(u, \phi) = K \exp\left(\frac{u}{\theta_2}\right) \int_u^\infty \exp\left(-w\phi\right) w^{n-1}(w - u)^{m-1} \, dw$$

$$= K \exp\left(\frac{u}{\theta_2}\right) A(u, \phi)$$

where (i) if $u < 0$

$$A(u, \phi) = \int_0^\infty \exp\left(-w\phi\right) w^{n-1}(w - u)^{m-1} \, dw$$

$$= \sum_{i=0}^{m-1} \binom{m-1}{i}(-u)^i \int_0^\infty \exp\left(-w\phi\right) w^{m+n-2-i} \, dw$$

$$= \frac{1}{\phi^{n+m-1}} \sum_{i=0}^{m-1} \binom{m-1}{i} \Gamma(n + m - 1 - i)(-u\phi)^i$$

and (ii) if $u > 0$

$$A(u, \phi) = \int_u^\infty \exp\left(-w\phi\right) w^{m-1}(w - u)^{m-1} \, dw$$

$$= \exp\left(-\phi u\right) \int_0^\infty \exp\left(-\phi v\right)(u + v)^{n-1} v^{m-1} \, dv$$

$$= \frac{\exp\left(-\phi u\right)}{\phi^{n+m-1}} \sum_{i=0}^{n-1} \binom{n-1}{i} \Gamma(n + m - 1 - i)(\phi u)^i$$

Hence, the conditional distribution of z_1 given $u = z_1 - z_2$ is given by

$$h(w \mid u, \phi) = \frac{\exp\left(-\phi w\right) w^{n-1}(w - u)^{m-1}}{A(u, \phi)} \qquad (3)$$

We can now use (3) to obtain confidence interval of confidence coefficients $(1 - \alpha)$ for $R_t(s, 2) = \exp\left(-t\phi\right)$, the reliability of a system with independent components having failure time distribution given by (1). We will illustrate the method by an example.

EXAMPLE 7.1   Suppose three units from population 1 and two units from population 2 failed at times (1.0, 1.4, 1.6) and (1 3, 1.7) hours where the populations refer to (1). We wish to find 90% confidence limits for the reliability of the system.

We have $z_1 = w = 4.0$, $z_2 = 3.0$, $u = 1.0 > 0$, $n + m = 5$

$$A(u, \phi) = \frac{\exp(-\phi u)}{\phi^4} \sum_0^2 \binom{2}{i} \Gamma(4 - i)(\phi u)^i$$

$$= \frac{\exp(-\phi u)}{\phi^4} (6 + 4\phi u + \phi^2 u^2)$$

$$h(w \mid u = 1, \phi) = C \exp[-\phi(w - 1)]w^2(w - 1)$$

where $C = \dfrac{\phi^4}{6 + 4\phi + \phi^2}$

Now

$$0.05 = P(w \leqslant 4)$$

$$= C \int_1^4 h(w \mid u = 1, \phi) \, dw$$

$$= C \int_0^3 \exp(-\phi y)(y^3 + 2y^2 + y) \, dy$$

$$= 1 - \frac{(6 + 22\phi + 40\phi^2 + 48\phi^3) \exp(-3\phi)}{6 + 4\phi + \phi^2}$$

We solve for

$$(6 + 22\phi + 40\phi^2 + 48\phi^3) \exp(-3\phi) = 0.95(6 + 4\phi + \phi^2)$$

and $\quad (6 + 22\phi + 40\phi^2 + 48\phi^3) \exp(-3\phi) = 0.05(6 + 4\phi + \phi^2)$

The solutions are $\phi_1 = 0.3804$, $\phi_2 = 2.2223$.

Since $R_{t_0}(s, 2) = \exp(-\phi t_0)$ is a decreasing function of ϕ, the two-sided 90% confidence bounds for R_{t_0} are given by

$$0.90 = P(\phi_1 < \phi < \phi_2)$$

$$= P[\exp(-\phi_2 t_0) < R_{t_0}(s, 2) < \exp(-\phi_1 t_0)]$$

$$= P[\exp(-2.2223 t_0) < R_{t_0}(s, 2) < \exp(-0.3804 t_0)].$$

We state here that Lentner-Buehler results can be used for failure censored samples where we have r failures of the type c_1 and s failures of the type c_2. Here we replace n, m, z_1 and z_2 by r, s, $\sum_1^r x_{(i)} + (n - r)x_{(r)}$ and $\sum_1^s y_{(i)} + (n - s)y_{(s)}$ respectively. It is quite obvious that Letner-Buehler method could be difficult even for the case $k = 2$ and small values of n and m such as 3 and 2 respectively used in Example 7.1. The main problem is that the solution of equations leading to ϕ_1 and ϕ_2 may be quite involved. For the case $k \geqslant 2$, Mawaziny and Buehler (1967) derived a method to give approximate solutions.

Lieberman and Ross (1971) presented a simple method for the related problems using different parametrization $\lambda_1 = 1/\theta_i$, $i = 1, 2$. Here $R_t(s, 2)$ $= \exp\{-(\lambda_1 + \lambda_2)t\}$. Lieberman and Ross proved that confidence interval of confidence coefficient $(1 - \alpha)$ has lower and upper limits $R_1^*(t)$ and $R_2^*(t)$ given by

$$R_1^*(t) = \exp\left[-\frac{t}{2u}\{\chi^2_{1-\alpha/2}(2k)\}\right] \tag{4}$$

$$R_2^*(t) = \exp\left[-\frac{t}{2u}\{\chi^2_{\alpha/2}(2k)\}\right] \tag{5}$$

where $\chi^2_p(2k)$ denotes the $100p\%$ point of a χ^2 r.v. with $2k$ degrees of freedom,

$$u = \text{Min}\left[\sum_1^n x_i, \sum_1^m y_i\right]$$

and

$$k = \left[\text{largest } j \leqslant n \,\bigg|\, \sum_1^j x_i \leqslant u\right] + \left[\text{largest } j \leqslant m \,\bigg|\, \sum_1^m y_i \leqslant u\right]$$

If the experiment is terminated at a pre-determined time t_c and n_1 out of n components from X and m_1 out of m components from Y failed at times $x_{(1)} < x_{(2)} < \ldots < x_{(n_1)} < t_c$ and $y_{(1)} < y_{(2)} < \ldots < y_{(m_1)} < t_c$ respectively, then Draper and Guttaman (1972) showed that Lieberman and Ross results can be modified as follows:

$$R_{c,1}^* = \exp\left[-\frac{\chi^2_{1-\alpha/2}(2k^*)}{2U^*}t_c\right] \tag{6}$$

$$R_{c,2}^* = \exp\left[-\frac{\chi^2_{\alpha/2}(2k^*)}{2U^*}t_c\right] \tag{7}$$

$$U^* = \text{Min}\left[\sum_1^{n_1} W_i, \sum_1^{m_1} Z_j\right]$$

$W_i = (n_1 - i + 1)(x_{(1)} - x_{(i-1)}), \quad i = 1, 2, \ldots, n_1; \; x_{(0)} = 0$

$Z_j = (m_1 - j + 1)(y_{(j)} - y_{(j-1)}), \; j = 1, 2, \ldots, m_1; \; y_{(0)} = 0$

$k^* = \left\{\text{largest } j \leqslant n_1 \,\bigg|\, \sum_1^j W_i \leqslant U^*\right\} + \left\{\text{largest } j \leqslant m_1 \,\bigg|\, \sum_1^j Z_i \leqslant U^*\right\}$

Sarkar (1971) presented a method for $k = 2$ and subsequently generalized it for censored samples from independent exponential distributions with parameters λ_1, λ_2. Let

$r = \text{Min}(n_1, m_1)$ and $v_i = (n - i + 1)(x_{(i)} - x_{(i-1)})$

$w_i = (m - i + 1)(y_{(i)} - y_{(i-1)}), \quad x_{(0)} = y_{(0)} = 0, \quad i = 1, 2, \ldots, r$

$z_i = \text{Min}(v_i, w_i)$

Since v_i and w_i are independent and exponentially distributed with parameters λ_1, λ_2,

$$1 - F(u) = P(z_i > u) = \exp(-\lambda u), \lambda = \lambda_1 + \lambda_2$$

Thus, $(z_1, z_2, \ldots, z_r)$ are independent identically distributed exponential variates with parameter λ.

Let $S_1(t)$, $S_2(t)$ be $(1 - \alpha)$-level lower and upper Sarkar's reliability bounds. Since $2\lambda \sum_1^r z_i \sim \chi^2(2r)$,

$$1 - \alpha = P[S_1(t) < R_t < S_2(t)]$$

$$= P\left[-\frac{1}{t} \log S_2(t) < \lambda < -\frac{1}{t} \log S_1(t) \right]$$

$$= P\left[-\frac{2 \sum_1^r z_i}{t} \log S_2(t) < \chi^2(2r) < -\frac{2 \sum_1^r z_i}{t} \log S_1(t) \right]$$

$$-\frac{2}{t} \sum_1^r z_i \log S_1(t) = \chi^2_{1-\alpha/2}(2r), \quad -\frac{2}{t} \sum_1^r z_i \log S_2(t) = \chi^2_{\alpha/2}(2r)$$

Hence,

$$S_1(t) = \exp\left[\frac{-t}{2 \sum_{i=1}^r z_i} \chi^2_{1-\alpha/2}(2r) \right] \tag{8}$$

$$S_2(t) = \exp\left[\frac{-t}{2 \sum_{i=1}^r z_i} \chi^2_{\alpha/2}(2r) \right] \tag{9}$$

We will illustrate these methods by two examples discussed by Draper and Guttman (1972).

EXAMPLE 7.2   Independent samples of size 3 and 4 are obtained on each of two components with the following results:

Components	Lifetime (in 1000 hours)			Sums
1	28.56	22.42	180.24	231.22
2	35.76	226.58	160.08 556.06	978.48

We have

$$\sum_{i=1}^4 x_i = 978.48, \quad \sum_{i=1}^3 y_i = 231.22$$

$$n = 4, \quad m = 3, \quad u = 231.22, \quad k = 4$$

For 90% confidence belt, $\chi^2_{.95}(8) = 15.507$, $\chi^2_{.05}(8) = 2.733$. Substituting in (4) and (5),

$$R_1^*(t) = \exp(-0.03353t), \quad R_2^*(t) = \exp(-0.005915t) \tag{10}$$

Functions (10) are plotted in Figure 1.

Substituting in (8) and (9),

$$S_1(t) = \exp(-0.0436t), \quad S_2(t) = \exp(-0.0056t) \tag{11}$$

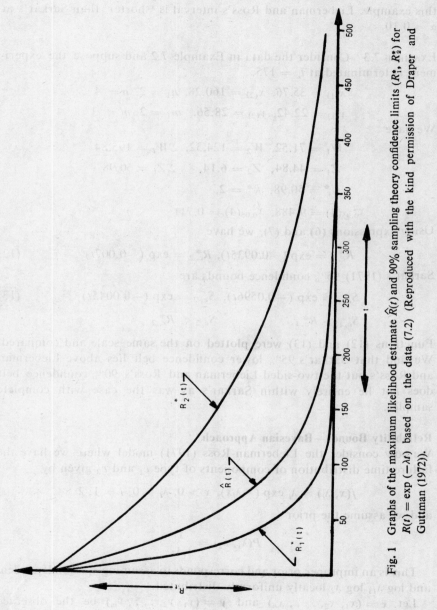

Fig. 1 Graphs of the maximum likelihood estimate $\hat{R}(t)$ and 90% sampling theory confidence limits (R_1^*, R_2^*) for $R(t) = \exp(-\lambda t)$ based on the data (7.2) (Reproduced with the kind permission of Draper and Guttman (1972)).

Clearly, $S_1(t) < R_1^*(t)$ and $S_2(t) < R_2^*(t)$. Functions (11) were plotted and the plots compared with Figure 1. It was found that Sarkar's 90% confidence belt contains Lieberman and Ross's 90% confidence belt, i.e., for this example, Lieberman and Ross's interval is 'shorter' than Sarkar's at $\alpha = 0.10$.

EXAMPLE 7.3   Consider the data in Example 7.2 and suppose the experiment is terminated at $t_c = 175$.

$$x_{(1)} = 35.76, \ x_{(2)} = 160.28, \ n_1 = 2, \ n = 4$$
$$y_{(1)} = 22.42, \ y_{(2)} = 28.56, \ m_1 = 2, \ m = 3$$

We have

$$W_1 = 71.52, \ W_2 = 124.32, \ \Sigma W_i = 195.84$$
$$Z_1 = 44.84, \ Z_2 = 6.14, \ \ \ \ \Sigma Z_i = 50.98$$
$$U_c^* = 50.98, \ k^* = 2,$$
$$\chi_{0.95}^2(4) = 9.488, \ \chi_{0.05}^2(4) = 0.711$$

Using expressions (6) and (7), we have

$$R_{c,1}^* = \exp(-0.0935t), \ R_{c,2}^* = \exp(-0.007t) \tag{12}$$

Sarkar's (1971) 95% confidence bounds are

$$S_{c,1} = \exp(-0.0596t), \ S_{c,2} = \exp(-0.0045t) \tag{13}$$
$$S_{c,1} > R_{c,1}^*, \ \ \ \ \ \ \ \ \ \ \ \ \ S_{c,2} < R_{c,2}^*$$

Functions (12) and (13) were plotted on the same scale and compared. We find that Sarkar's 95% lower confidence belt lies above Lieberman and Ross's but the two-sided Lieberman and Ross's 90% confidence belt does not lie entirely within Sarkar's as was the case with complete samples.

Reliability Bounds—Bayesian Approach

We now consider the Lieberman-Ross (1971) model where we have the failure time distribution of components of type c_1 and c_2 given by

$$f(x, \lambda_i) = \lambda_i \exp(-\lambda_i x), \ x > 0, \ \lambda_i > 0, \ i = 1, 2$$

and we assume the prior

$$P(\lambda_1, \lambda_2) = \frac{1}{\lambda_1 \lambda_2}$$

This is an improper prior and corresponds to assuming λ_1, λ_2 independent and $\log \lambda_1, \log \lambda_2$ locally uniformly distributed.

Let $x = (x_1, x_2, \ldots, x_n)$ and $y = (y_1, y_2, \ldots, y_m)$ be the observed failure times of the components of type c_1 and c_2 respectively and let

$$T = \sum_1^n x_i, \ S = \sum_1^m y_i$$

The usual computations lead to the posterior distribution of (λ_1, λ_2) given by

$$\Pi(\lambda_1, \lambda_2 \mid x, y) = K \exp(-\lambda_1 T - \lambda_2 S)\lambda_1^{n-1}\lambda_2^{m-1} \tag{14}$$

where

$$K = \frac{T^n S^m}{\Gamma(n)\Gamma(m)}$$

which implies that given (T, S), λ_1, λ_2 are independent gamma variates with shape parameters n and m and scale parameters $1/T$, $1/S$ respectively. Bayes estimator of $R_t(s, 2) = \exp\{-(\lambda_1 + \lambda_2)\}$ is given by

$$\hat{R}_t^{(B)} = E[R_t(s, 2) \mid x, y]$$

$$= \int_0^\infty \int_0^\infty \exp\{-(\lambda_1 + \lambda_2)t\}\Pi(\lambda_1, \lambda_2 \mid x, y) \, d\lambda_1 \, d\lambda_2$$

$$= \left(1 + \frac{t}{T}\right)^{-n}\left(1 + \frac{t}{S}\right)^{-m} \tag{15}$$

Note that for large n and m the Bayes estimator $\hat{R}_t^{(B)}$ will be very close to the MLE $\hat{R}(t) = \exp\left\{-t\left(\frac{1}{x} + \frac{1}{y}\right)\right\}$ and it can be shown that $|\hat{R}_t^{(B)} - \hat{R}(t)|$ converges in probability to zero as $n \to \infty$ and $m \to \infty$. Since $\lambda = \lambda_1 + \lambda_2$ is the parameter of interest we obtain the posterior distribution of λ given x and y. We use the transformation $\lambda_1 = \lambda u$ and $\lambda_2 = \lambda(1 - u)$.

The Jacobian of transformation is $|J| = \lambda$. Substituting in (14) we obtain

$$\Pi(\lambda, u \mid x, y) = K\lambda^{m+n-1} \exp(-\lambda S) \exp\{-\lambda u(T - S)\}u^{n-1}(1 - u)^{m-1}$$

Integrating out u, the posterior distribution of λ is given by

$$\Pi(\lambda \mid x, y) = K\lambda^{m+n-1} \exp(-\lambda S) \int_0^1 \exp\{-\lambda u(T - S)\}u^{n-1}(1 - u)^{m-1} \, du$$

The explicit evaluation of this distribution is difficult. However, a programme has been developed (Disenhouse and Ghezzo, 1972) whereby one can solve for λ_α defined by the equation

$$\alpha = \int_{\lambda_\alpha}^\infty \Pi(\lambda \mid x, y) \, d\lambda \tag{16}$$

and use it to obtain the one or two-sided confidence limits for $R_t(s, 2)$ $= \exp(-\lambda t)$.

Suppose we want to obtain the lower confidence limit $R_1(t)$ such that

$$1 - \alpha = P[R_1(t) < R_t(s, 2) = \exp(-t\lambda_\alpha)]$$

$$= P\left[\lambda < \frac{1}{t} \log \frac{1}{R_1(t)}\right]$$

or equivalently

$$\alpha = P\left[\lambda \geqslant \frac{1}{t} \log \frac{1}{R_1(t)} = \lambda_\alpha\right]$$

Thus, $R_1(t) = \exp(-t\lambda_\alpha)$ where λ_α has been defined in (16). Similarly, two-sided confidence limits for $R_t(s, 2)$ would be

$$R_1(t) = \exp(-t\lambda_{\alpha/2}) \quad \text{and} \quad R_2(t) = \exp(-t\lambda_{1-\alpha/2}) \tag{17}$$

EXAMPLE 7.4   Consider the data in Example 7.2. The maximum likelihood and the Bayes estimators of reliability are given by

$$\widehat{R}_t = \exp\left\{-t\left(\frac{4}{978.48} + \frac{3}{231.22}\right)\right\} = \exp(-0.0175t)$$

$$\widehat{R}_t^B = \left(1 + \frac{t}{978.48}\right)^{-4}\left(1 + \frac{t}{231.22}\right)^{-3}$$

Using the programme already referred to, $\lambda_{0.05} = 0.03160$ and $\lambda_{0.95} = 0.00695$ and the 95% lower posterior confidence limit of R_t is $\exp(-0.03160t)$. Substituting in (17), the 90% posterior confidence limits of R_t are $R_1(t) = \exp(-0.03160t)$ and $R_2(t) = \exp(-0.00695t)$.

$$\{R_1^*(t), \widehat{R}(t), R_2^*(t)\} \quad \text{and} \quad \{R_1(t), \widehat{R}_t^B, R_2(t)\}$$

are plotted in Figures 1 and 2 on the same scale. We find

(i) $\widehat{R}(t) \leqslant \widehat{R}_t^B$,

(ii) $R_1^*(t) \leqslant R_1(t)$ and $R_2^*(t) \geqslant R_2(t)$.

Thus, for this example, Bayes interval is 'shorter' than the sampling counterpart (Draper and Guttman, 1972).

Parallel System

Suppose we have a system made up of k components $(c_1, c_2, \ldots, c_k)$ connected in parallel. Let the life time X of each component be independently and identically distributed as $f(x, \theta) = \frac{1}{\theta}\exp\left(-\frac{x}{\theta}\right)$, $x > 0$, $\theta > 0$.

As defined in section 1, the system reliability is given by

$$\begin{aligned} R_t(p, k) &= 1 - P(c_1 \leqslant t, c_2 \leqslant t, \ldots c_k \leqslant t) \\ &= 1 - (1 - R_t)^k \\ &= \sum_{j=1}^{k}(-1)^{j-1}\binom{k}{j}R_t^j \\ &= \sum_{j=1}^{k}(-1)^{j-1}\binom{k}{j}\exp\left(-\frac{jt}{\theta}\right). \end{aligned} \tag{18}$$

If the life times are independently but not identically distributed, we have

$$R_t(p, k) = 1 - \prod_{i=1}^{k}\left\{1 - \exp\left(\frac{t}{\theta_i}\right)\right\}, \text{ where } \theta_i = E(X_i) \tag{19}$$

We note that, in general, the reliability of a parallel system is higher than the reliability of a series system with the same number of similar components. This follows from the results discussed in section 1 which

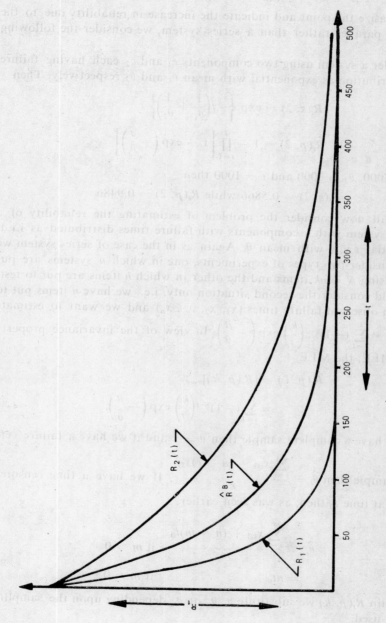

Fig. 2   Graphs of the Bayes estimate $\widehat{R}_B(t)$ and 90% posterior confidence limits (R_1, R_2) for $R(t) = \exp(-\lambda)t$, based on the data (7.2) (Reproduced with the kind permission of Draper and Guttman (1972)).

imply that

$$R_t(s, k) \leqslant \underset{1 \leqslant i \leqslant k}{\text{Min }} R_i(t) \leqslant \underset{1 \leqslant i \leqslant k}{\text{Max }} R_i(t) \leqslant R_t(p, k)$$

To emphasize the point and indicate the increase in reliability due to the use of a parallel rather than a series system, we consider the following example.

Consider a system using two components c_1 and c_2 each having failure time distribution as exponential with mean θ_1 and θ_2 respectively. Then

$$R_t(s, 2) = \exp\left\{-t\left(\frac{1}{\theta_1} + \frac{1}{\theta_2}\right)\right\}$$

$$R_t(p, 2) = 1 - \prod_{i=1}^{2}\left[1 - \exp\left(-\frac{t}{\theta_i}\right)\right]$$

If $\theta_1 = 5000$, $\theta_2 = 3000$ and $t = 1000$ then

$$R_t(s, 2) = 0.5866 \text{ while } R_t(p, 2) = 0.9486$$

We will now consider the problem of estimating the reliability of a parallel system with k components with failure times distributed as i.i.d. exponential r.v.'s with mean θ. Again as in the case of series system we could consider two types of experiments, one in which n systems are put to test using $N = nk$ items and the other in which n items are put to test. We would consider the second situation only, i.e., we have n items put to test with observed failure times $(x_1, x_2, \ldots x_n)$ and we want to estimate $R_t(p, k) = \sum_{j=1}^{k} (-1)^{k-j}\binom{k}{j} \exp\left(-\frac{jt}{\theta}\right)$. In view of the invariance property of the MLE, the MLE

$$\hat{R}_t(p, k) = [R_t(p, k)]_{\theta=\hat{\theta}}$$

$$= \sum_{j=1}^{k} (-1)^{j-1}\binom{k}{j} \exp\left(-\frac{jt}{\hat{\theta}}\right) \tag{20}$$

If we have a complete sample then $\hat{\theta} = \bar{x}$ and if we have a failure censored sample then $\hat{\theta}_r = \dfrac{\sum_{i=1}^{r} x_{(i)} + (n - r)x_{(r)}}{r}$. If we have a time censored sample at time t_0 then, as was seen earlier,

$$\hat{\theta}_{t_0} = \frac{\sum_{i=1}^{m} x_{(i)} + (n - m)t_0}{m} \qquad \text{if } m > 0$$

$$= nt_0 \qquad\qquad\qquad \text{if } m = 0$$

To obtain $\hat{R}_t(p, k)$ we substitute $\hat{\theta}$, $\hat{\theta}_r$ or $\hat{\theta}_{t_0}$ depending upon the sampling scheme used.

To obtain UMVUE of $R_t(p, k)$ we note that if we obtain UMVUE of $\exp(-jt/\theta)$ for $j = 1, 2, \ldots k$, then we can obtain the UMVUE of $R_t(p, k)$ by substituting these estimates in the expression (18). For details we refer to Rutemiller (1966). The UMVUE for the complete sample is

$$\tilde{R}_t(p, k) = \sum_{j=1}^{k} (-1)^{j-1} \binom{k}{j} \left(1 - \frac{jt}{S_n}\right)^{n-1} \quad \text{if } S_n \geqslant jt, j = 1, 2, \ldots k$$

$$\text{and } S_n = \sum_{i=1}^{n} x_i$$

Note that if $S_n < jt$ for $j = j_0$ then $S_n < jt$ for $j \geqslant j_0$ and the UMVUE of $\exp(-jt/\theta)$ is zero. Therefore in general

$$\tilde{R}_t(p, k) = \sum_{j=1}^{k} (-1)^{j-1} \binom{k}{j} \left[\left(1 - \frac{jt}{S_n}\right)^+ \right]^{n-1}$$

where $\left(1 - \dfrac{jt}{S_n}\right)^+ = 1 - \dfrac{jt}{S_n} \quad \text{if } S_n \geqslant jt$

$$= 0 \qquad \text{otherwise}$$

In case we have a failure censored sample, we obtain the UMVUE $\tilde{R}_t(p, k)$ by replacing n by r and using $S_r = \sum_{i=1}^{k} x_{(i)} + (n - r)x_{(r)}$ instead of S_n. For the time truncated sample we note that the problem of obtaining UMVUE of $R_t(p, k)$ is not easy in that UMVUE for the case $k = 1$ is not available and hence the problem for $k > 1$ is very complicated.

One can combine series and parallel systems in many interesting ways and the computation of reliability for such systems can be done by using the basic formulae given in section 1. For example, let us consider two systems S_1 and S_2 connected in parallel where S_1 consists of k_1 identical components connected in series and S_2 consists of k_2 identical components connected in series. To obtain the reliability of such a system $R_t(S_1 \cup S_2, k_1, k_2)$ we note that $R_t(S_1 \cup S_2, k_1, k_2) = 1 - [1 - R_t(S_1, k_1)][1 - R_t(S_2, k_2)]$ since S_1 and S_2 are connected in parallel. However $R_t(S_1, k_1) = [R_1(t)]^{k_1}$ and $R_2(S_2, k_2) = [R_2(t)]^{k_2}$ and thus,

$$R_t(S_1 \cup S_2, k_1, k_2) = 1 - [1 - R_1^k(t)][1 - R_2^k(t)] \tag{21}$$

If S_1 and S_2 are connected in series rather than in parallel then the reliability of such a system is given by

$$R_t(S_1 \cap S_2, k_1, k_2) = R_1^{k_1}(t) \cdot R_2^{k_2}(t) \tag{22}$$

If the components of S_1 and S_2 are exponential r.v.'s with means θ_1 and θ_2 respectively then

$$R_t(S_1 \cup S_2, k_1, k_2) = 1 - \left[1 - \exp\left(-\frac{tk_1}{\theta_1}\right)\right]\left[1 - \exp\left(-\frac{tk_2}{\theta_2}\right)\right]$$

$$R_t(S_1 \cap S_2, k_1, k_2) = \exp\left\{-t\left(\frac{k_1}{\theta_1} + \frac{k_2}{\theta_2}\right)\right\}$$

We will not go into the problems of estimation of reliability of such complex systems, as this takes the matter very much outside the scope of the text.

EXERCISES

7.1 Independent samples of size 6 and 8 are obtained on each of the two components with the following results.

Components	Life time (in hours)			Sums
1	30.48	25.63	29.34	445.34
	45.49	152.67	161.34	
2	48.96	54.67	100.75	1411.10
	120.68	148.34	228.43	
	293.45	415.82		

Compute Lieberman and Ross (1971), 90% confidence belt for R_t at any time t.

7.2 Repeat the problem 7.1 assuming that the test is terminated at $t_c = 150$ hours.

7.3 Assuming Lieberman and Ross (1971) model (in series) in problem 7.1, compute
(a) the MLE of R_t.
(b) Bayes estimator of R_t under a prior $p(\lambda_1, \lambda_2) = $ constant, at $t = 100$ hours.

7.4 Suppose the life time X of each component is independent and identically distributed as one-parameter exponential with mean life θ.

Let $n = 15$, $k = 3$, $t = 5$, $S_{15} = 20$

Obtain the MLE and UMVUE of $R_t(p, 3)$.

7.5 Let the failure time distribution be given by

$$f(x, \theta) = \frac{1}{\theta^\lambda \Gamma(\lambda)} x^{\lambda-1} \exp\left(-\frac{x}{\theta}\right), \quad x > 0, \theta > 0$$

where λ is known. Given a complete sample $(x_1, x_2, \ldots x_n)$, the UMVUE of $R_t(p, k)$ is given by

$$\widetilde{R}_t(p, k) = \Gamma(n\lambda)\{\Gamma(\lambda)\}^k [\Gamma\{(n-k)\lambda\}]^{-1}, \, T^{1-n\lambda} \sum_{j=1}^{k} (-1)^{j-1} \binom{k}{j}$$

$$\times \int \int \ldots \int (x_1 x_2 \ldots x_k)^{p-1} (T - x_1 - x_2 - \ldots x_k)^{(n-k)p-1} \, dx_1 \, dx_2 \ldots dx_k$$

the region of integration being defined by the conditions $x_i \geqslant t, i = 1, 2, \ldots, k$ with

$$x_1 + x_2 + \ldots + x_k \leqslant T = \sum_{i=1}^{n} x_i.$$

Obtain $\widetilde{R}_t(p, k)$, given $\lambda = 2$, $n = 4$, $k = 2$, $T = 3$, $t = 1$ (Wani and Kabe, 1971).

7.6 Putting $\lambda = 1$ in 7.5, we have the one-parameter exponential failure time distribution

$$f(x, \theta) = \frac{1}{\theta} \exp\left(-x/\theta\right), \quad x > 0, \theta > 0$$

Using the results given by Wani and Kabe (1971) obtain $\widetilde{R}_t(p, k)$, given $n = 4$, $k = 2$, $T = 3$, $t = 1$.

Gamma and Beta Distributions

A r.v. X with p.d.f. given by $f(x, a) = \dfrac{1}{\Gamma(a)} e^{-x} x^{a-1}$, $x > 0$, $a > 0$ is defined to be a gamma variate (r.v.) with the parameter a. The function $\Gamma(a) = \displaystyle\int_0^\infty e^{-x} x^{a-1} dx$ for $a > 0$ is known as the gamma function and the corresponding integral as the Eulerian integral. The gamma function occurs in various contexts in applied mathematics. We refer to Abramowitz and Stegun (1964) for various formulae involving gamma functions and also for extensive tables of gamma function. We now first establish the well-known recurrence relation that

$$\Gamma(a) = (a-1)\Gamma(a-1), \quad a > 1$$

$$\Gamma(a) = \int_0^\infty e^{-x} x^{a-1} \, dx$$

$$= \left[e^{-x} x^{a-1}(-1) \right]_0^\infty + (a-1) \int_0^\infty x^{a-2} e^{-x} \, dx$$

$$= (a-1)\Gamma(a-1), \quad \text{since } a > 1$$

If $a >$ some positive integer r, then we can continue to integrate by parts to obtain successively

$$\Gamma(a) = (a-1)\Gamma(a-1)$$
$$= (a-1)(a-2)\Gamma(a-2)$$
$$= (a-1)(a-2)\ldots(a-r)\Gamma(a-r)$$

In particular if a itself is integer ($a > 1$), by taking $r = a - 1$ we obtain

$$\Gamma(a) = (a-1)(a-2)\ldots 1 \cdot \Gamma(1) = (a-1)! \text{ since } \Gamma(1) = \int_0^\infty e^{-x} \, dx = 1.$$

Using these formulae it is easy to check that if X is a gamma variate with parameter a, then

$$E(X) = \frac{\Gamma(a+1)}{\Gamma(a)} = a, \ E(X^2) = \frac{\Gamma(a+2)}{\Gamma(a)} = (a+1)a \text{ and Var } (X) = a$$

Consider $Y = X\sigma$ where X is a gamma variate with parameter p; then the p.d.f. of Y is given by

$$f(y, \sigma, p) = \frac{1}{\Gamma(p)} e^{-y/\sigma} \frac{y^{p-1}}{\sigma^p}, \quad y > 0, \sigma > 0, p > 0$$

For $p = n/2$ and $\sigma = 2$ we get the well-known χ_n^2 distribution which has been used extensively in the text. Thus if $Y = \chi_n^2$, then $Y/2$ is a gamma variate with parameter $p = n/2$.

Next, consider two independent gamma random variables X, Y with parameters a and b respectively. Then the joint p.d.f. of (X, Y) is given by:

$$f(x, y) = \frac{1}{\Gamma(a)\Gamma(b)} \, e^{-(x+y)} x^{a-1} y^{b-1}, \; x > 0, \, y > 0$$

We will consider the r.v. $U = X/(X + Y)$ and obtain its p.d.f. This we obtain by using the joint transformation $u = x/(x + y)$ and $v = x$. This gives $x = v$ and $y = v(1 - u)/u$. The range of (u, v) is $\{0 < u < 1, 0 < v < \infty\}$ and the Jacobian of the transformation is $|J| = v/u^2$. Thus, the joint p.d.f. of (U, V) is now given by

$$f(u, v) = \frac{1}{\Gamma(a)\Gamma(b)} \exp\left(-\frac{v}{u}\right) v^{a-1} \frac{v^{(b-1)}(1 - u)^{b-1} v}{u^{b-1} u^2}$$

$$= \frac{1}{\Gamma(a)\Gamma(b)} \exp\left(-\frac{v}{u}\right) v^{a+b-1} \frac{(1 - u)^{b-1}}{u^{b+1}}, \quad 0 < v < \infty, \, 0 < u < 1$$

To obtain the p.d.f. of U, we integrate out v and the marginal p.d.f. of U is given by

$$f_1(u) = \int_0^\infty f(u, v) \, dv$$

$$= \frac{(1 - u)^{b-1}}{u^{b+1}} \int_0^\infty \exp\left(-\frac{v}{u}\right) v^{a+b-1} \, dv \left[\frac{1}{\Gamma(a)\Gamma(b)}\right]$$

$$= \frac{(1 - u)^{b-1}}{u^{b+1}} \frac{\Gamma(a + b)}{(1/u)^{a+b}} \frac{1}{\Gamma(a)\Gamma(b)}$$

$$= \frac{\Gamma(a + b)}{\Gamma(a)\Gamma(b)} u^{a-1}(1 - u)^{b-1}, \quad 0 < u < 1$$

The function $\left[\dfrac{\Gamma(a + b)}{\Gamma(a)\Gamma(b)}\right]^{-1} = B(a, b)$ is the well-known beta function which again has been well tabulated for various positive values of a and b. Both beta and gamma functions have been studied extensively and were first discovered by the famous mathematician Euler and are also known as Eulerian functions. The associated integrals $\int_0^\infty e^{-x} x^{a-1} \, dx = \Gamma(a)$ and $B(a, b) = \int_0^1 x^{a-1}(1 - x)^{b-1} \, dx$ for $a > 0$, $b > 0$, are known as Eulerian integrals.

The corresponding distribution function

$$P(X \leqslant x) = \frac{1}{\Gamma(a)} \int_0^x e^{-w} w^{a-1} \, dw = I_a(x)$$

is known as the incomplete gamma function and is also available in tabulated form for selected combinations of values of a and x. Similarly

$$P(U \leqslant x) = B_x(a, b) = \frac{1}{B(a, b)} \int_0^x u^{a-1}(1 - u)^{b-1} \, du$$

is known as incomplete Beta function and has been tabulated for selected combinations of values of (a, b) and x. The r.v. U with p.d.f.

$$f_1(u) = \frac{1}{B(a, b)} u^{a-1}(1 - u)^{b-1}, \, 0 < u < 1$$

is referred to as a beta random variable of the first kind with parameters a and b. It is easy to show that $E(U) = a/(a + b)$ and $\text{Var}(U) = a/(a + b - 1)(a + b)^2$ and this problem is left to the reader as an exercise.

Again consider (X, Y) independent gamma r.v.'s with parameters (a, b). Consider the r.v. $V = X/Y$. We obtain the marginal p.d.f. of V by using the transformation $v = x/y$, $w = y$. This gives $x = vw$ and $y = w$. The range of (v, w) is $\{0 < v < \infty, 0 < w < \infty\}$ and the Jacobian of the transformation is $|J| = w$. Therefore, the joint p.d.f. of (V, W) is given by

$$f(v, w) = \frac{e^{-w(v+1)}v^{a-1}w^{a+b-1}}{\Gamma(a)\Gamma(b)}, \, 0 < u < \infty, 0 < w < \infty$$

The marginal of V is obtained by integrating out W and is given by

$$\begin{aligned} f_1(v) &= \left[\int_0^\infty e^{-w(v+1)}w^{a+b-1} \, dw \right] \frac{v^{a-1}}{\Gamma(a)\Gamma(b)} \\ &= \frac{\Gamma(a+b)}{\Gamma(a)\Gamma(b)} \frac{v^{a-1}}{(1 + v)^{a+b}} \\ &= \frac{1}{B(a, b)} \frac{v^{a-1}}{(1 + v)^{a+b}}, \, 0 < v < \infty \end{aligned}$$

The r.v. V is referred to as beta variate (r.v.) of the second kind and is connected with the well-known F distribution used extensively in statistics. We will consider the relationship in some detail. Let X be a χ_m^2 r.v., i.e., X is gamma with $p = m/2$ and $\sigma = 2$. Similarly let Y be a χ_n^2 r.v., i.e., Y is gamma with $p = n/2$ and $\sigma = 2$. The F ratio, or the F r.v. with m and n degrees of freedom is defined by

$$F_{m, n} = \frac{\dfrac{X_m'}{m}}{\dfrac{Y_n'}{n}} = \frac{n}{m} \frac{\chi_m^2}{\chi_n^2},$$

$$F_{m, n} = \frac{n}{m} \frac{X'}{Y'}$$

where $X' = X/\sigma$ and $Y' = Y/\sigma$ are gamma r.v.'s with $p = m/2$ and $p = n/2$ respectively. X and Y are assumed to be independent and so are X' and Y'.

Thus $(m/n)F_{m, n} = X'/Y'$ is distributed as V with $a = m/2$, $b = n/2$. Hence $F_{m, n}$ has density given by

$$h(F_{m, n}) = \frac{1}{B\left(\dfrac{m}{2}, \dfrac{n}{2}\right)} \left(\frac{m}{n}\right)^{m/2} \frac{F^{(m/2)-1}}{\left(1 + \dfrac{m}{n}F\right)^{(m+n)/2}}, \, F > 0$$

This is the F r.v. with m and n degrees of freedom. Again extensive tables are available for the d.f. of the $F_{m,n}$ r.v. For tables related to gamma and beta r.v.'s as well as F distribution we refer to Tables for Incomplete Beta and Gamma functions, K. Pearson (1968), Biometrika Tables for Statisticians (Pearson and Hartley, 1954) and also the well-known tables of Fisher and Yates (1957).

We now prove the additive property of the gamma r.v.'s, i.e., if X and Y are independent gamma r.v.'s with parameters a and b respectively, then $(X + Y)$ is gamma with parameter $(a + b)$.

Consider the joint p.d.f. of (X, Y) given by

$$f(x, y) = \frac{1}{\Gamma(a)\Gamma(b)}\, e^{-(x+y)}x^{a-1}y^{b-1}, \quad 0 < x < \infty, 0 < y < \infty$$

Consider the transformation $x + y = v$ and $u = x$; then the range of (x, y) goes into $\{0 < u < v < \infty\}$ and the $|J| = 1$. Thus the joint p.d.f. of (U, V) is given by

$$f(u, v) = \frac{1}{\Gamma(a)\Gamma(b)}\, e^{-v}u^{a-1}(v - u)^{b-1}, \quad 0 < u < v < \infty$$

The marginal p.d.f. of V is obtained by integrating out U, thus

$$\begin{aligned}
f_1(v) &= \frac{e^{-v}}{\Gamma(a)\Gamma(b)} \int_0^v u^{a-1}(v - u)^{b-1}\, du \\
&= \frac{e^{-v}v^{a+b-1}}{\Gamma(a)\Gamma(b)} \int_0^1 t^{a-1}(1 - t)^{b-1}\, dt \\
&= \frac{e^{-v}v^{a+b-1}}{\Gamma(a + b)}, \quad 0 < v < \infty
\end{aligned}$$

Here we have used the transformation $u = vt$.

By using induction, we can show that if $(X_1, X_2, \ldots, X_n)$ are independent gamma r.v.'s with parameters $(a_1, a_2, \ldots, a_n)$ respectively, then $(X_1 + X_2 + \ldots + X_n)$ is a gamma with parameter $(a_1 + a_2 + \ldots + a_n)$.

Order Statistics

Let $(X_1, X_2, \ldots, X_n)$ be the random sample with realized values $X_1 = x_1$, $X_2 = x_2, \ldots, X_n = x_n$. We assume that $(X_1, X_2, \ldots, X_n)$ are i.i.d.r.v.'s with p.d.f. $f(x)$. Since the r.v.'s are assumed continuous, the probability that two observations are equal is zero. Thus, we can arrange the observations in increasing order of magnitude $x_{(1)} < x_{(2)} < \ldots < x_{(n)}$. The corresponding r.v.'s $\{X_{(1)}, X_{(2)}, \ldots, X_{(n)}\}$ are called the order statistics of the sample and $X_{(r)}$ is called the rth component of the order statistics or the rth order statistic. Clearly we have $X_{(1)} < X_{(2)} < \ldots < X_{(n)}$ and the probability that any two or more components of the order statistics are equal, is zero. It may be noted that in actual practice because of the limits to accuracy of the measurement, two observations can be equal. Thus, according to our assumptions the probability that two items put on test fail at the same instant of time is zero but their recorded lives in practice could turn out to be equal, say to the nearest second. If such a tie occurs among the observations, we break the tie by interpolating for the next place. Thus, if two failure times are recorded as 601 hours then we know that observations are recorded correct to the nearest hour. The two observations at 601 hours are now recorded as 601.24 and 600.76 say. Note that the total is not altered.

We will now first obtain the d.f. of $X_{(r)}$. Let

$$
\begin{aligned}
G_r(x) &= P\{X_{(r)} \leqslant x\} \\
&= P\{\text{at least } r \text{ observations} \leqslant x\} \\
&= \sum_{k=r}^{n} \binom{n}{k} [F(x)]^k [1 - F(x)]^{n-k}
\end{aligned}
$$

since the probability that any observation $X_i < x$ is $F(x)$ and the probability that exactly k observations are less than x is given by

$$
\binom{n}{k} [F(x)]^k [1 - F(x)]^{n-k}
$$

By differentiating $G_r(x)$ we can obtain the p.d.f. of $X_{(r)}$ given by $g_r(x)$, where

$$
g_r(x) = \frac{n!}{(r-1)! \, (n-r)!} [F(x)]^{r-1} [1 - F(x)]^{n-r} f(x), \quad -\infty < x < \infty
$$

We will obtain the density of $X_{(r)}$, by using the method of differentials. First we note that

$$
\begin{aligned}
P\{x - \delta x \leqslant x_i \leqslant x + \delta x\} &= F(x + \delta x) - F(x - \delta x) \\
&= f(x) \, \delta x + 0(\delta x)
\end{aligned}
$$

where $0(\delta x)$ are terms which, when divided by δx, go to zero as $\delta x \to 0$. Thus,

$$
g_r(x) \, \delta x + 0(\delta x) = P\{x - \delta x \leqslant X_{(r)} \leqslant x + \delta x\}
$$

Further, the probability that more than one observation X_i, X_j both belong to $(x - \delta x, x + \delta x) = [f(x) \, \delta x + 0(\delta x)]^2 = 0(\delta x)$.

As the following figure shows, except for terms $0(\delta x)$, the following configuration gives us the event $\{x - \delta x \leqslant X_{(r)} < x + \delta x\}$.

$$
\begin{array}{ccc}
& (r-1)\ X_i\text{'s} \qquad \text{one } X_i \qquad (n-r)\ X_i\text{'s} & \\
-\infty \overline{ \underset{x - \delta x}{|} \underset{x + \delta x}{|} } +\infty
\end{array}
$$

Therefore,

$$
\begin{aligned}
g_r(x) \, \delta x + 0(\delta x) &= P[(r-1) \text{ observations} < x - \delta x, \text{ exactly one} \\
&\quad \text{observation in } (x - \delta x, x + \delta x) \text{ and } (n-r) \\
&\quad \text{observations} > x + \delta x] + 0(\delta x) \\
&= \frac{n!}{(r-1)! \, (n-r)! \, 1!} [F(x - \delta x)]^{r-1} [F(x + \delta x) \\
&\quad - F(x - \delta x)][1 - F(x + \delta x)]^{n-r} + 0(\delta x)
\end{aligned}
$$

Dividing both the sides by δx and taking limits as $\delta x \to 0$, we have

$$g_r(x) = \frac{n!}{(r-1)! \, (n-r)!} [F(x)]^{r-1}[1 - F(x)]^{n-r}f(x), \quad -\infty < x < \infty$$

In the same manner we can obtain the joint p.d.f. of $X_{(r)}$ and $X_{(s)}$, the rth and sth components of order statistics where we assume $r < s$. Let $g_{r,s}(u, v)$ denote the joint p.d.f. of $\{X_{(r)}, X_{(s)}\}$. Then

$$g_{r,s}(u, v) \, \delta u \, \delta v + 0(\delta u \delta v)$$

is the probability that

$$\{(u - \delta u \leqslant X_{(r)} \leqslant u + \delta u), (v - \delta v \leqslant X_{(s)} \leqslant v + \delta v)\}$$

Clearly, $u + \delta u < v - \delta v$, since $X_{(r)} < X_{(s)}$. Again except for terms of $0(\delta u \, \delta v)$, the following configuration gives us the event

$$\{u - \delta u \leqslant X_{(r)} \leqslant u + \delta u < v - \delta v \leqslant X_{(s)} \leqslant v + \delta v\}$$

Therefore,

$$\begin{aligned}
g_{r,s}(u,v)uv + 0(\delta u \, \delta v) = {} & \frac{n!}{(r-1)! \, (s-r-1)! \, (n-s)!} [F(u - \delta u)]^{r-1} \\
& \times [f(u) \, \delta u][F(v + \delta v) - F(u + \delta u)]^{s-r-1} \\
& \times [f(v) \, \delta v][1 - F(v + \delta v)]^{n-s} + 0(\delta u \, \delta v)
\end{aligned}$$

Note that as $\delta u \to 0$, $\delta v \to 0$, $F(u + \delta u) \to F(u)$ and $F(v + \delta v) \to F(v)$. Therefore, dividing both sides by $(\delta u \, \delta v)$ and then taking limits, as $\delta u \to 0$, $\delta v \to 0$,

$$\begin{aligned}
g_{r,s}(u, v) = {} & \frac{n!}{(r-1)! \, (s-r-1)! \, (n-s)!} [F(u)]^{r-1}[F(v) - F(u)]^{s-r-1} \\
& \times [1 - F(u)]^{n-s}f(u)f(v), \quad -\infty < u < v < +\infty
\end{aligned}$$

We can use the same argument to obtain the joint p.d.f. of all the order statistics $\{X_{(1)}, X_{(2)}, \ldots, X_{(n)}\}$ given by $g(x_{(1)}, x_{(2)}, \ldots, x_{(n)})$. Again the probability that

$$\{x_{(i)} - \delta x_{(i)} \leqslant X_{(i)} < x_{(i)} + \delta x_{(i)}, \, i = 1, 2, \ldots, n\}$$

is given by

$$\begin{aligned}
g(x_{(1)}, x_{(2)}, \ldots, x_{(n)}) \, \delta x_{(1)} \ldots, \delta x_{(n)} + 0(\delta x_{(1)}, \ldots, \delta x_{(n)}) \\
= n! \, f(x_{(1)}) \, \delta x_{(1)} f(x_{(2)}) \, \delta x_{(2)} \ldots f(x_{(n)}) \, \delta x_{(n)} + 0(\delta x_{(1)} \, \delta x_{(2)} \ldots \delta x_{(n)})
\end{aligned}$$

Dividing by $\delta x_{(1)} \, \delta x_{(2)} \ldots \delta x_{(n)}$ and taking limits as $\delta x_{(r)} \to 0$ for $r = 1, 2, \ldots, n$, we have

$$\begin{aligned}
g(x_{(1)}, x_{(2)}, \ldots, x_{(n)}) = n! \, f(x_{(1)}) \, f(x_{(2)}) \ldots f(x_{(n)}), \\
-\infty < x_{(1)} < x_{(2)} < \ldots < x_{(n)} < \infty
\end{aligned}$$

The marginal p.d.f. of $x_{(r)}$ can be obtained by integrating out $\{x_{(1)}, \ldots x_{(r-1)}$ and $x_{(r+1)}, \ldots x_{(n)}\}$ over the above region for fixed $x_{(r)}$. Similarly the marginal p.d.f. of $X_{(r)}$, $X_{(s)}$ can also be obtained by integrating out the remaining $X_{(i)}$ variables for fixed $x_{(r)}$ and $x_{(s)}$. This is left to the reader as an exercise.

Theory of Estimation and Tests of Hypotheses

In this section we will outline some of the basic results in parametric inference which we have frequently used in the text. Let $(X_1, X_2, \ldots, X_n)$ be a random sample of size n from a population with p.d.f. $f(x, \theta)$ where f is known and θ is an unknown constant in that we do not know the exact value of θ but only know the possible range of values of θ given by a set Ω. θ is called the parameter and Ω is called the parameter space. We assume further that θ is a labelling parameter in that if we know θ say $\theta = \theta_0$, then the p.d.f. is completely known and the population is identified and two different values of θ say θ_1 and θ_2 cannot lead to the same p.d.f. The objective is to estimate or test hypotheses about the indexing parameter θ or some function of θ say $\psi(\theta)$. We will first consider the estimation problem.

An estimator is a function of observations only given by say $T(X_1, \ldots, X_n)$ which we employ to estimate $\psi(\theta)$ on the basis of a sample $(X_1, X_2, \ldots, X_n)$. The estimate T (or the value of a statistic T) may thus vary from sample to sample and T is in fact a random variable. How do we evaluate the suitability of T as an estimator of $\psi(\theta)$?

Note that $T(X_1, X_2, \ldots, X_n)$ is a r.v. and has a d.f. given by $G(t, \theta) = P[T \leqslant t \,|\, \theta]$ and associated p.d.f. $g(t, \theta)$ which effectively depends on θ. Thus, if population is normal with mean μ and variance σ^2 and $\bar{X}_n$, the sample mean is the statistic under consideration, then we know that $\bar{X}_n$ is itself normal with mean μ and variance σ^2/n. The effectiveness of T as an estimator of $\psi(\theta)$ would thus depend upon the properties of the class of distribution functions $\{G(t, \theta), \theta \in \Omega\}$ or the corresponding class of p.d.f.s $\{g(t, \theta), \theta \in \Omega\}$.

An estimator T is said to be unbiased for $\psi(\theta)$ if $E(T \,|\, \theta) = \psi(\theta)$ holds for every $\theta \in \Omega$. An unbiased estimator, thus, has its distribution centered at $\psi(\theta)$, the function of the parameter which we want to estimate. If $E(T \,|\, \theta) \neq \psi(\theta)$ for some $\theta \in \Omega$ then T is said to be a biased estimator for $\psi(\theta)$ and $E(T \,|\, \theta) - \psi(\theta) = b(\theta)$ denotes the bias of T. We generally prefer unbiased estimator to a biased one. Usually there are several unbiased estimators of $\psi(\theta)$ and we may have to choose one within this class. This choice is made by comparing the variances of the unbiased estimators, or by looking at the second moment of the distributions of the unbiased estimators. Let T_1 and T_2 be both unbiased for $\psi(\theta)$, then we prefer T_1 to T_2 if var $(T_1 \mid \theta) \leqslant$ var $(T_2 \mid \theta)$ for every $\theta \in \Omega$ with strict inequality for some $\theta_0 \in \Omega$. An estimator T_0 is called Uniformly Minimum Variance Unbiased Estimator (UMVUE) of $\psi(\theta)$ if T_0 is unbiased for $\psi(\theta)$ and for any other unbiased estimator T we prefer T_0 to T. Recall that the variance is a measure of dispersion of the distribution and the smaller the variance the more concentrated is the distribution around the mean. Thus UMVUE T_0 of $\psi(\theta)$ has its distribution as much concentrated around $\psi(\theta)$ as is possible and this holds for every value of $\theta \in \Omega$.

How does one obtain a UMVUE for $\psi(\theta)$ in a given model $\{f(x, \theta), \theta \in \Omega\}$? We assume that U_ψ, the class of all unbiased estimators of $\psi(\theta)$ is not empty—and indeed in general U_ψ contains more than one element. The well-known theorems of Rao-Blackwell and Lehman-Scheffé give a method in the case where the model under consideration admits a complete sufficient statistic. Let $\{L(X, \theta), \theta \in \Omega\}$ be the likelihood of the sample. Let $T(X)$ be a statistic (real or vector valued) with the associated class of p.d.f.s $\{g(t, \theta), \theta \in \Omega\}$. Then T is said to be sufficient for the family $\{L(X, \theta), \theta \in \Omega\}$ (or equivalently sufficient for θ) if

$$L(X, \theta) = g(T(X), \theta)h(X)$$

where $h(X)$ is a function of X only and does not depend on θ. One can interpret $h(X)$ as the conditional p.d.f. of the observations X given $T(X)$, and T is sufficient if this conditional p.d.f. is independent of θ. The concept of the sufficient statistic is very important and sufficient statistic plays a very important role in any problem of statistical inference. We now state Rao-Blackwell theorem.

Rao-Blackwell Theorem

Let T be sufficient for θ and let T_1 be an unbiased estimator of $\psi(\theta)$ such that Var $(T_1 \mid \theta)$ is finite. Then $E(T_1 \mid T = t) = u(t)$ is independent of θ and

  (i) $E[u(T) \mid \theta] = \psi(\theta)$

  (ii) Var $[u(T) \mid \theta] \leqslant$ Var $[T_1 \mid \theta]$.

Thus, any unbiased estimator T_1 of $\psi(\theta)$ can be improved by taking conditional expectation, $u(T) = E(T_1 \mid T)$, since the variance of $u(T)$ is always smaller than that of T_1, and $u(T) \in U_\psi$. Thus, the search for UMVUE can be restricted to estimators which are functions of sufficient statistics. For the proof we refer to the texts of Hogg and Craig (1978), or Mood, Graybill and Boes (1974).

Now suppose that we have two different unbiased estimators T_1 and T_2 of $\psi(\theta)$ and by Rao-Blackwellization, i.e., conditioning w.r.t. the sufficient statistic T, we obtain $u_1(T)$ and $u_2(T)$ both unbiased for ψ and both having smaller variance than Var $(T_1 \mid \theta)$. How does one choose between $u_1(T)$ and $u_2(T)$? In one fortunate situation this problem does not arise in that $u_1(T)$ and $u_2(T)$ are identical. In fact in this situation there is a unique unbiased estimator which is a function of sufficient statistic T. This case would occur if the sufficient statistic T is such that the class of p.d.f.s $\{g(t, \theta), \theta \in \Omega\}$ is also complete. A class of p.d.f.s $\{g(t, \theta), \theta \in \Omega\}$ is said to be complete if $E[\phi(T) \mid \theta] = 0$ for every $\theta \in \Omega$ implies that $\phi(T) = 0$. Thus, the only unbiased estimator of the constant zero is the function which is identically equal to zero. If $\{g(t, \theta), \theta \in \Omega\}$ is a complete class of p.d.f.s (usually paraphrased as if T is complete) then $\psi(\theta)$ has a unique unbiased estimator $u(T)$. For if not then let $u_1(T)$ and $u_2(T)$ be both unbiased for $\psi(\theta)$. Then clearly $\phi(T) = u_1(T) - u_2(T)$ is unbiased for zero and in view of completeness of T, $u_1(T) - u_2(T) = 0$ or $u_1(T) = u_2(T)$. This leads us to the Rao-Blackwell-Lehman-Scheffé theorem.

Rao-Blackwell-Lehman-Scheffé Theorem

If T is complete and sufficient for θ, then any function $u(T)$ is UMVUE of its expectation.

We note that the property of completeness of the sufficient statistic is a delicate property and proofs are difficult and depend heavily on the theory of Laplace-Transforms and analytic functions of complex variables. The above theorems also hold when the statistic T and the parameter θ are both vector valued. For details and proofs we refer the reader to Lehman (1959).

In statistical inference problems the large sample theory plays a very important role and defines some general methods of estimation such as Maximum Likelihood and the Method of Moments. Here it is assumed that the samples are fairly large and estimators are studied from the viewpoint of the properties in the limit, as n, the sample size, tends to infinity. So it is now assumed that $\{X_i\}^n$ are i.i.d. r.v.'s with p.d.f. $\{f(X, \theta), \theta \in \Omega\}$ and n is large and $T^{(n)}(X_1, X_2, \ldots, X_n)$ is a sequence of estimators with associated p.d.f.s $\{g_n(t, \theta), \theta \in \Omega\}$ and we study the behaviour of $T^{(n)}$ as $n \to \infty$.

Let us consider the case when $T^{(n)}$ is to be used as an estimator of θ. Now $(T^{(n)} - \theta)$ is regarded as an error and it is desired that $(T^{(n)} - \theta)$ should be small in large samples at least. However, since $(T^{(n)} - \theta)$ is a r.v. we cannot generally guarantee this for every sample and we must consider the $P\{|T^{(n)} - \theta| < \epsilon \mid \theta\} = p_n(\epsilon, \theta)$. If we can show that $\lim_{n \to \infty} p_n(\epsilon, \theta) = 1$ for every $\epsilon > 0$ and each $\theta \in \Omega$ then T_n is said to be a consistent estimator of θ, and this fact is expressed by saying that $T^{(n)}$ converges stochastically or in probability to θ and the notation used is $T^{(n)} \overset{p}{\to} \theta$ as $n \to \infty$.

We are generally interested in not just the stochastic convergence of $T^{(n)}$ to θ as $n \to \infty$ but also the rate at which $T^{(n)} \overset{p}{\to} \theta$ or $p_n(\epsilon, \theta) \to 1$ as $n \to \infty$. By Tchebychev's inequality this rate is connected with MSE $(T^{(n)} \mid \theta)$, the mean squared error of $T^{(n)}$ and as in the UMVUE approach we search for consistent estimators $T^{(n)}$ such that MSE $(T^{(n)} \mid \theta) \to 0$ as $n \to \infty$ as fast as possible. Of course in general we are interested in the coverage probability $p_n(\epsilon, \theta)$ and this can also be studied using asymptotic or large sample distribution of $T^{(n)}$. This leads to the Consistent Asymptotic Normal (CAN) estimators. An estimator $T^{(n)}$ is said to be a CAN for θ if $\sqrt{n}(T^{(n)} - \theta)$ converges in distribution to $N(0, \sigma_T^2(\theta))$. For more details on stochastic convergence and convergence in distribution we refer the readers to Hogg and Craig (1978) and Mood, Graybill and Boes (1974). As in the UMVUE approach, here the search is for the Best Asymptotically Normal (BAN) estimator, essentially that estimator which has the smallest possible (asymptotic) variance $\sigma_T^2(\theta)/n$. The general theory of the CAN and BAN estimators is rather complex and interested students are referred to Zacks (1971). A property enjoyed by consistent as well as CAN estimator is invariance under the functional transformation. Let $T^{(n)}$ be consistent for θ and let g be a continuous function, then $g(T^{(n)})$ is consistent for $g(\theta)$ and if $T^{(n)}$ is CAN for θ with g a differentiable function with $dg/d\theta \neq 0$, then $g(T^{(n)})$ is CAN for $g(\theta)$ with the asymptotic variance given by $\sigma_T^2(\theta) \, (dg/d\theta)^2/n$. Note that such a property is not enjoyed by the unbiased estimators. If T is unbiased for θ then in general $g(T)$ is not unbiased for $g(\theta)$, except when g is linear.

There are several methods for generating CAN estimators and the ones most often used are the Method of Moments and the Method of Maximum Likelihood. In the Method of Moments we select a suitable function $T(X)$ of the observed r.v. X such that $E[T(X) \mid \theta] = \psi(\theta)$ and Var $(T \mid \theta) < \infty$ and these are well behaved functions of θ. Thus, we require ψ to be such that ψ^{-1} is well defined and $\left(\dfrac{d\psi^{-1}}{d\theta}\right) = \dfrac{1}{d\psi/d\theta} \neq 0$. Then the Method of Moments consists in equating the sample mean $\dfrac{1}{n} \sum_{i=1}^{n} T(X_i)$ with its expectation namely $\psi(\theta)$ to give a solution $\tilde{\theta} = \psi^{-1}\left[\dfrac{1}{n} \sum_{i=1}^{n} T(X_i)\right]$. Then it can be shown that $\tilde{\theta}$ is CAN for θ with the asymptotic variance Var $(T \mid \theta) \dfrac{1}{n} \left[\dfrac{d\psi}{d\theta}\right]^{-2}$. The function $T(X)$ is generally chosen in such a way that $\dfrac{1}{n} \sum_{i=1}^{n} T(X_i)$ is sufficient for θ. The Method of Maximum Likelihood consists in finding out that value $\hat{\theta} \in \Omega$ such that $L(X \mid \theta)$, for fixed X and variations in $\theta \in \Omega$, is maximized at $\hat{\theta}$. The MLE $\hat{\theta}$ can be shown to be CAN under suitable regularity conditions which are too complex to list here. A general method to obtain MLE is to solve the likelihood equation $(d \log L)/d\theta = 0$ and determine $\hat{\theta}$ so that there is a (relative) maximum at $\hat{\theta}$. If $L(X \mid \theta)$ is sufficiently well behaved, i.e., the range is independent of parameter and $\log L$ is at least twice differentiable in θ, then solving $(d \log L)/d\theta = 0$ leads to $\hat{\theta}$ a CAN estimator with the asymptotic variance $1/nI(\theta)$ where $nI(\theta) = E(d \log L/d\theta)^2 = E(-d^2 \log L/d\theta^2)$ is the Fisher information index. For specific regularity conditions and details we refer to Zacks (1971) and Wilks (1962).

Many of these results can be extended to the case when θ is vector valued. Thus, an estimator $T = (T_1, T_2, \ldots, T_k)$ is consistent for $\theta = (\theta_1, \theta_2, \ldots, \theta_k)$ if $T_j \xrightarrow{P} \theta_j$, $j = 1, 2, \ldots, k$ i.e., T_j is consistent for θ_j for each $j = 1, 2, \ldots, k$. Similarly, T is said to be CAN if $\sqrt{n}(T - \theta)$ converges in distribution to the k variate normal distribution with mean vector zero and (non-singular) variance covariance matrix $\Sigma = \|\sigma_{ij}\|$ where $\sigma_{ij} = $ covariance (T_i, T_j). The Method of Maximum Likelihood then leads to $\hat{\theta} = (\hat{\theta}_1, \hat{\theta}_2, \ldots, \hat{\theta}_k)$ which in general can be obtained by solving the set of likelihood equations $(\partial \log L)/\partial \theta_i = 0$, $i = 1, 2, \ldots, k$. It can be shown that under certain regularity conditions, the asymptotic variance co-variance matrix of $(\hat{\theta} - \theta)$ is given by the inverse of the Fisher Information matrix $nI(\theta) = \|nI_{ij}(\theta)\|$ where

$$nI_{ij}(\theta) = E\left(\frac{\partial \log L}{\partial \theta_i} \cdot \frac{\partial \log L}{\partial \theta_j}\right) = E\left(-\frac{\partial^2 \log L}{\partial \theta_i \, \partial \theta_j}\right)$$

Similarly, for the Method of Moments, we construct k moment equations for the k unknown parameters based on a suitable k dimensional statistic $T = (T_1, T_2, \ldots, T_k)$ such that $E(T_i \mid \theta) = \psi_i(\theta_1, \theta_2, \ldots, \theta_k)$ $i = 1, 2, \ldots, k$ and $(\psi_1, \psi_2, \ldots, \psi_k)$ are differentiable functions of θ and the variance covariance matrix of $(T_1, T_2, \ldots, T_k)$ is given by $\Sigma = \|\sigma_{ij}(\theta)\|$. The moment equations then are $\frac{1}{n}\sum_{j=1}^{n} T_i(X_j) = \psi_i(\theta_1, \theta_2, \ldots, \theta_k)$, $i = 1, 2, \ldots, k$ which leads to the moment estimators $(\tilde{\theta}_1, \tilde{\theta}_2, \ldots, \tilde{\theta}_k)$ provided the Jacobian $\left|\frac{\partial(\psi_1, \psi_2, \ldots, \psi_k)}{\partial(\theta_1, \theta_2, \ldots, \theta_k)}\right| = |\Delta|$ is non-zero where $\Delta_{ij} = \frac{\partial \psi_i(\theta_1, \ldots, \theta_k)}{\partial \theta_j}$. We can show that $(\tilde{\theta}_1, \tilde{\theta}_2, \ldots, \tilde{\theta}_k)$ are CAN for $(\theta_1, \theta_2, \ldots, \theta_k)$ in that $\{\sqrt{n}(\tilde{\theta}_1 - \theta_1), \sqrt{n}(\tilde{\theta}_2, - \theta_2), \ldots, \sqrt{n}(\tilde{\theta}_k - \theta_k)\}$ has in the limit a k variate normal distribution with mean vector zero and the variance covariance matrix $\Delta^{-1}\Sigma\Delta$.

We now consider the problems of testing the hypotheses. Here the interest lies not in estimating the individual value of the labelling parameter θ but to ascertain whether θ is in a certain subregion of Ω, the parameter space under consideration. This situation has already been illustrated in Chapter 5 by a suitable example. Let Ω be divided into two parts Ω_{H_0} and Ω_{H_1} which are mutually disjoint and $\Omega_{H_0} \cup \Omega_{H_1} = \Omega$. The statement $\theta \in \Omega_{H_0}$ will be referred to as the null hypothesis and $\theta \in \Omega_{H_1}$ as the alternative hypothesis. On the basis of a random sample $x = (x_1, x_2, \ldots, x_n)$ we have to decide whether the null hypothesis H_0 or the alternative hypothesis H_1 holds. This is generally done by dividing the sample space $\mathscr{X}$ into two parts, say A and its complement A^c, where A is called the Acceptance Region (AR) and A^c is called the Critical Region (CR). If the observed sample $x \in A$ then we assert $\theta \in \Omega_{H_0}$ or accept H_0 and if $x \notin A$ or $x \in A^c$ then we assert that $\theta \in \Omega_{H_1}$ or reject H_0. Note that if we accept H_0 then we reject H_1 and if we accept H_1 then we reject H_0. There is, thus, a duality and the procedure for accepting or rejecting H_0 (or H_1) can be specified either by specifying A or A^c. It is traditional to describe a test procedure by the critical region A^c rather than the acceptance region A. Let W_0 denote the CR defining the corresponding test procedure ψ_0. Clearly there are many CRs available and we must evaluate the effectiveness of performance of these CRs. This evaluation is done in terms of the two errors associated with the CR, W_0 or the corresponding test procedure ψ_0. The first error consists in rejecting H_0 (accepting H_1) when in fact H_0 holds (or $\theta \in \Omega_{H_0}$) and the other error consists in accepting H_0 (rejecting H_1) when H_1 holds. Since the observed x is a r.v. these errors are also random and therefore we consider the probabilities of occurrence of these errors. Thus, the type I error is measured by the probability of rejecting H_0 when H_0 holds and the type II error is measured by the probability of accepting H_0 when H_1 holds. A reasonable test procedure would have both, type I and type II errors small and the best procedure would be the one for which these errors are as small as possible. However, the 'best' is impossible since to decrease the type I error we must remove some sample points x from the CR W_0 and, therefore, add these to W_0^c (AR). We can reduce the type I error only at the cost of increasing the type II error and some compromise is necessary.

Let us consider the simplest case where Ω consists of two points only and $\Omega_{H_0} = \{\theta_0\}$ and $\Omega_{H_1} = \{\theta_1\}$. Thus, we have to decide between two competing models with p.d.f.s that can be denoted f_0 and f_1. Let W be any CR, then we have

$$\text{Type I Error} = P\,[\text{rejecting } H_0 \mid H_0 \text{ true}] = \int_W \cdots \int \left[\prod_{i=1}^n f_0(x_i) \right] dx_i$$

$$\text{Type II Error} = P\,[\text{accepting } H_0 \mid H_1 \text{ true}] = \int_{W^c} \cdots \int \prod_{i=1}^n f_1(x_i)\, dx_i$$

The type I error associated with the critical region W is called the size or the level of significance of W. Instead of considering the type II error which depends on acceptance region W^c, it is customary to consider the probability of the complementary event, the probability of rejecting H_0 when H_1 is true. This is associated with critical region W, and is called the power of the critical region W. Clearly, power $= 1 -$ type II error and the optimal procedure would be the one which minimizes the size and maximizes the power. Let

$$\alpha(W) = \int_W \cdots \int \prod_{i=1}^n f_0(x_i)\, dx_i = \text{size of } W$$

$$\beta(W) = \int_W \cdots \int \prod_{i=1}^n f_1(x_i)\, dx_i = \text{power of } W$$

Note that $\alpha(W)$ and $\beta(W)$ are increasing functions of W and as such we cannot minimize $\alpha(W)$ and maximize $\beta(W)$ simultaneously. A compromise then is made by the convention that the type I error (or size) $\alpha(W)$ must be controlled with an upper limit α and subject to this restriction, we maximize $\beta(W)$. Thus, the best critical region is W_0 such that $\alpha(W_0) \leqslant \alpha$ and for any critical region W satisfying $\alpha(W) \leqslant \alpha$ we have $\beta(W_0) \geqslant \beta(W)$.

For discriminating between the two simple hypotheses H_0 and H_1 each consisting of a single completely specified distribution the best critical region of (a given) size exists and is given by the well known Neyman-Pearson lemma, by

$$W_0 = \left\{ x \,\middle|\, \prod_{i=1}^{n} \frac{f_0(x_i)}{f_1(x_i)} \leqslant k \right\}$$

where k is a constant to be chosen such that $P[x \in W_0 \mid H_0] = \alpha$. For the proof we refer to Hogg and Craig (1978).

Next consider the problem when the null hypothesis is simple and $\Omega_{H_0} = \{\theta_0\}$, but the alternatives are composite, i.e., Ω_{H_1} consists of more than one θ say for example $\theta > \theta_0$. Let $\beta(\theta) = P\{\text{rejecting } H_0 \mid \theta\}$. Then $\beta_W(\theta)$ is called the power function of the critical region W. Again we are interested in a critical region W_0 for which $\beta_{W_0}(\theta_0)$ is as small as possible but $\beta_{W_0}(\theta)$ is as large as possible for each $\theta \in \Omega_{H_1}$. Note that $\beta_{W_0}(\theta_0)$ is the size of the critical region W_0 and $\beta_{W_0}(\theta_1)$ is the power of the critical region W_0 at the point $\theta_1 \in \Omega_{H_1}$. Since we know how to obtain the best critical region for testing a simple null against a simple alternative, let us take a specific point $\theta_1 \in \Omega_{H_1}$ and try to obtain a most powerful test (MPT) for the subproblem $H_0 : \theta = \theta_0$ vs. $H_1' : \theta > \theta_0$. By the Neyman-Pearson lemma the best critical region corresponding to the most powerful test is given by

$$W_{\theta_1} = \left\{ x \,\middle|\, \prod_{i=1}^{n} \frac{f(x_i, \theta_0)}{f(x_i, \theta_1)} \leqslant k \right\}$$

and the constant k would generally depend upon the level of significance α, and the particular alternative θ_1 considered in the subproblem. In the fortunate situation where k depends only on α but not on θ_1, the best critical region for the subproblem would be independent of particular $\theta_1 \in \Omega_{H_1}$ and the same critical region $W_{\theta_1} \equiv W_0$ serves as the critical region for the original (larger problem) of testing $H_0 : \theta = \theta_0$ against $H_1 : \theta \in \Omega_{H_1}$. For example, the reader can verify that for the case where $f(x, \theta)$ is normal with mean θ and variance unity for the problem $H_0 : \theta = 0$ vs. $H_1 : \theta > 0$ we have $W_{\theta_1} = \left\{ x \mid \bar{x} \geqslant \frac{\xi_{1-\alpha}}{\sqrt{n}} \right\}$ where $\xi_{1-\alpha}$ is $100(1 - \alpha)\%$ point of the standard normal d.f. Clearly W_{θ_1} is independent of the particular θ_1 chosen from the set of alternatives $\theta > 0$ and $W_0 \equiv W_{\theta_1}$ is the best critical region and we have the solution for the original problem.

On the other hand, consider the problem of testing $H_0 : \theta = 0$ vs. $H_1 : \theta \neq 0$. Then if we select $\theta = \theta_1 < 0$ from the set of alternatives $W_{\theta_1} = \left\{ x \mid \bar{x} \leqslant \frac{\xi_\alpha}{\sqrt{n}} \right\}$, but if $\theta_1 > 0$ then $W_{\theta_1} = \left\{ x \mid \bar{x} \geqslant \frac{\xi_{1-\alpha}}{\sqrt{n}} \right\}$. Thus, the critical region depends upon the sign (though not the magnitude) of the particular alternative selected from Ω_{H_1}. There is, therefore, no best critical region or the uniformly most powerful test, for the problem $H_0 : \theta = 0$

vs. $H_1 : \theta \neq 0$. This is a common feature in two-sided alternatives and some compromise again has to be made. The problem becomes much harder when we are testing composite null against composite alternatives, i.e., both Ω_{H_0} and Ω_{H_1} contain many values of θ. Thus, for example in $N(\theta, 1)$ model we may want to test $H_0 : a \leqslant \theta \leqslant b$ while the alternatives specified by $H_1 : \theta < a$ or $\theta > b$. Similarly, in $N(\theta, \sigma^2)$ model we may want to test $H_0 : \theta = 0$ against $H_1 : \theta > 0$ with σ^2 unspecified under H_0 as well as H_1. In the latter case σ^2 is called a nuisance parameter.

Suppose now that Ω_{H_0} and Ω_{H_1} are both composite. Let W be any critical region and let

$$\beta_W(\theta) = P[x \in W \mid \theta]$$

Then W_0 is the best critical region if

$$\beta_{W_0}(\theta) \leqslant \alpha \text{ for any } \theta \in \Omega_{H_0}$$

and if

$$\beta_{W_0}(\theta) \geqslant \beta_W(\theta) \text{ for any } \theta \in \Omega_{H_1}$$

where W satisfies the size condition, viz., $\beta_W(\theta) \leqslant \alpha$ for any $\theta \in \Omega_{H_0}$. Such best critical regions rarely exist when Ω_{H_0} and Ω_{H_1} are both composite.

In order to give an acceptable solution some compromises are essential. Two such compromises are recommended which are based on the concepts of unbiasedness and similarity. A critical region is called unbiased if its power is always larger than the level of significance. Thus, W is unbiased if

$$\underset{\theta \in \Omega_{H_1}}{\text{Min }} \{\beta_W(\theta)\} \geqslant \underset{\theta \in \Omega_{H_0}}{\text{Max }} \{\beta_W(\theta)\}$$

Intuitively this condition is meaningful in that for a biased critical region, the power is less than the level of significance or P [rejecting $H_0 \mid H_0$ is false] is smaller than P [rejecting $H_0 \mid H_0$ is true], which is rather undesirable. We, thus, require that critical region be unbiased and then within the class of unbiased critical regions try to select one which is uniformly most powerful. In case of $N(\theta, 1)$ model with $H_0 : \theta = 0$ against $H_1 : \theta \neq 0$ we can show that the two-sided critical region $W_0 = \left\{ x \mid |\bar{x}| \geqslant \dfrac{\xi_{1-\alpha}}{\sqrt{n}} \right\}$ is uniformly most powerful within the class of unbiased critical regions. The criteria of unbiasedness is invoked when in general there are no nuisance parameters involved.

In case of problems where Ω_{H_0} involves a nuisance parameter, the property of similarity is suggested as a desirable property. For example in the $N(\theta, \sigma^2)$ model, suppose we want to test $H_0 : \theta = 0$ against $H_1 : \theta > 0$ and σ^2 unspecified, so that σ^2 acts as a nuisance parameter. Consider a critical region W and

$$\beta_W(\sigma^2) = P[x \in W \mid \theta = 0, \sigma^2]$$

the level of significance of W or size of W. This would generally depend on the nuisance parameter σ^2. If there is a critical region W such that $\beta_W(\sigma^2)$ is independent of σ^2, the nuisance parameter, then such a critical region is called a similar region. In general W is a similar (critical) region of size α if $\beta_W(\theta) = \alpha$ for every $\theta \in \Omega_{H_0}$.

Having decided to use the property of similarity, one then tries to obtain the best critical region (maximum possible power) within the class of similar regions. The general problem of the existence as well as the construction of the best critical region within the class of unbiased and/or similar regions is extremely complicated and reference may be made to Lehmann (1959) and Wilks (1962).

A very general method, known as the generalized likelihood ratio test most often leads to the best critical region for the problem $\theta \in \Omega_{H_0}$ against $\theta \in \Omega_{H_1}$. Let $\Omega = \Omega_{H_0} \cup \Omega_{H_1}$. The likelihood ratio is defined as

$$\lambda(x) = \sup_{\theta \in \Omega_{H_0}} \{L(x \mid \theta)\} / \sup_{\theta \in \Omega} \{L(x \mid \theta)\}$$

Note that for x discrete, $\sup L(x \mid \theta)$ denotes the maximum possible probability of obtaining the data x if in fact $\theta \in \Omega_{H_0}$ or H_0 is true. This then is compared with the maximum possible probability of obtaining data x under either H_0 or H_1. Clearly $0 < \lambda(x) \leqslant 1$ and values of $\lambda(x)$ near one support H_0 or indicate $\theta \in \Omega_{H_0}$ while small values of $\lambda(x)$ near zero support H_1 or indicate $\theta \in \Omega_{H_1}$. Hence, the critical region for the likelihood ratio test method is given by $W_0 = \{x \mid \lambda(x) \leqslant k\}$. In many instances, as is shown in normal or exponential models, the critical region W_0 is also equivalent to $W_0 = \{x \mid T(x) \leqslant k'\}$ where $T(x)$ is a suitable test statistic whose distribution under H_0 is known. Using the distribution of T under H_0, the constant k' is determined so that the size condition is satisfied. If there are situations in which we cannot find equivalent description of the critical region W_0 based on such a $T(x)$ then we must use $\lambda(x)$ itself. However, the exact distribution of $\lambda(x)$ is very difficult to determine and only asymptotic (large sample) distribution of $-2 \log \lambda(x)$ is known. It was proved by Wilks (1962) that for large samples $-2 \log \lambda(x)$ is distributed as a χ_r^2, where r, the degrees of freedom, is given by the difference between the number of independent parameters specified by Ω and Ω_{H_0}. For details we refer to Wilks (1962), as well as Rohatgi (1976). This result is only asymptotic and should be used only in case the critical region $\{x \mid \lambda(x) \leqslant k\}$ cannot be reduced to the one based on a statistic $T(x)$ whose (asymptotic) distribution is known.

References

Abramowitz, M. and Stegun, I. (Ed). (1964), *Handbook of Mathematical Functions*, National Bureau of Standards, Applied Mathematical Series 55, U.S. Government Printing Office.

Aitchison, J. and Brown, J. A. C. (1957), *The Log-normal Distribution*, Cambridge University Press.

Bain, L. J. and Antle, C. E. (1967), "Estimation of parameters in the Weibull distribution", *Technometrics* 9, 621–627.

Bain, L. J. and Weeks, D. L. (1965), "Tolerance limits for the generalized gamma distribution", *J. Amer. Statist. Assoc.* 60, 1142–1152.

Barnett, V. (1973), *Comparative Statistical Inference*, Wiley.

Bartholomew, D. J. (1957), "A problem in life testing", *J. Amer. Statist. Assoc.* 65, 350–355.

————— (1963), "The sampling distribution of an estimate arising in life testing", *Technometrics* 5, 361–374.

Bartlett, M. S. (1953), "Approximate confidence intervals", *Biometrika* 40, 12–19.

Basu, A. P. (1964), "Estimates of reliability for some distributions useful in life testing", *Technometrics* 6, 215–219.

Bayes, T. (1763), "An essay towards solving a problem in the doctrine of chances", *Phil. Trans. Roy. Soc.* 53, 370–418. Reprinted in *Biometrika* 45, 296–315.

Bazovsky, I. (1961), *Reliability Theory and Practice*, Prentice Hall, New Jersey.

Bhattacharya, S. K. (1967), "Bayesian approach to life testing and reliability estimation", *J. Amer. Statist. Assoc.* 26, 48–62.

Birnbaum, Z. W. and Saunders, S. C. (1958), "A statistical model for life lengths of material", *J. Amer. Statist. Assoc.* 53, 151–160.

Boardman, T. J. and Kendell, P. J. (1970), "Estimation in compound exponential failure models", *Tehnometrics* 2, 891–900.

Boardman, T. J. (1973), "Estimation in compound exponential failure models—when the data are grouped", *Technometrics* 15, 271–277.

Bogdanoff, D. A. and Pierce, D. A. (1973), "Bayes Fiducial inference for the Weibull distribution", *J. Amer. Statist. Assoc.* 68, 659–664.

Box, G. E. P. and Tiao, G. C. (1973), *Bayesian Inference in Statistical Analysis*, Addison-Wesley.

Brownlee, K. A. (1949), Industrial Experimentation, 2nd ed., H.M.S.O. London.

Chapman, D. G. (1956), "Estimating the parameters of a truncated gamma distribution", *Ann. Math. Statist.* 27, 498–506.

Cohen, Jr., A. C. (1950), "Estimating parameters of Pearson type III populations from truncated samples", *J. Amer. Statist. Assoc.* 45, 411–423.

————— (1965), "Maximum likelihood estimation with Weibull distribution based on complete and on censored samples", *Technometrics* 7, 579–588.

————— (1967), "Estimation in mixtures of two normal distributions", *Technometrics* 9, 15–25.

Cramér, H. (1961), *Mathematical Methods of Statistics*, Princeton University Press.

David, H. A. (1970), *Order Statistics*, Wiley.

Davis. D. J. (1952), "The analysis of some failure data", *J. Amer. Statist. Assoc.* **47**, 113–150.

Deemer, Jr., W. L. and Votaw, Jr. D. F. (1955), "Estimation of parameters of truncated or censored exponential distributions", *Ann. Math. Statist.* **26**, 498–504.

Delaporte, P. (1950), "Etude Statistique sur les proprietes des fontes", *Res. Inst. Int. Statist.* **18**, 161.

Den Broeder, G. G. (1955), "On parameter estimation for truncated Pearson type III distributions", *Ann. Math. Statist.* **26**, 659–663.

Des Raj (1953), "Estimation of the parameters of type III populations from truncated samples", *J. Amer. Statist. Assoc.* **48**, 336–349.

Dick, N. P. and Bowden, D. C. (1973), "Maximum likelihood estimation for mixtures of two normal population distributions", *Biometrics* **29**, 781–790.

Disenhouse, J. and Ghezzo, H. (1972), See Draper, N. R. and Guttman, I. (1972).

Draper, N. R. and Guttman, I. (1965), "Transformation of life-test data", *Technical Report No. 60*, Department of Statistics, University of Wisconsin.

————— (1972), "The reliability of independent exponential series systems—a Bayesian approach", *Technical Report No. 317*, Department of Statistics, University of Wisconsin.

Epstein, B. (1947), "The mathematical description of certain breakage mechanism leading to the logarithmico-normal distribution", *J. Franklin Inst.* **224**, 471.

————— (1958), "Exponential distribution and its role in life testing", *Industrial Quality Control*, 4-9.

————— (1960), "Test for the validity of the assumption that the underlying distribution of life is exponential", *Technometrics* **2**, 83–101, 167–183.

————— (1961), "Estimation of bounded relative error for the mean life of exponential distribution" *Technometrics* **61**, 107.

Erdélyi, A. (1954), *Tables of Integral Transformations, Vol. 1*, McGraw-Hill.

Fisher, R. A. and Yates, F. (1957), *Statistical Tables for Biological, Agricultural and Medical Research*, Hafner Publishing, New York.

Fraser, D. A. S. (1958), *Statistics, an Introduction*, Wiley.

Giesbrecht, F. and Kempthorne, O. (1976), "Maximum likelihood estimation with three parameter lognormal distribution", *J. Roy. Statist. Soc.* **38** (3), 257–264.

Gnedenko, B. V., Belyayev, Yu, K. and Solovyev, A. D. (1969), *Mathematical Models of Reliability*, Academic Press.

Greenwood, J. A. and Durand, D. (1960), "Aids for fitting the gamma distribution by maximum likelihood", *Technometrics* **2**, 55–65.

Grubbs, F. E. (1971), "Approximate fiducial bounds on reliability for the two-parameter negative exponential distribution", *Technometrics* **13**, 873–876.

Gupta, A. K. (1952), "Estimation of the mean and the standard deviation of a normal population from a censored sample", *Biometrika* **39**, 260-273.

Gupta, S. S. (1960), "Order statistics from the gamma distribution", *Technometrics* **2**, 243-262.

Gupta, S. S. and Groll, P. A. (1961), "Gamma distribution in acceptance sampling based on life tests", *J. Amer. Statist. Assoc.* **56**, 942-970.

Harter, H. L. (1961), "Estimating the parameters of the negative exponential population using one or two order statistics", *Ann. Math. Statist.* **32**, 1078-1084.

Harter, H. L. and Moore, A. H. (1965), "Point and interval estimation, based on order statistics, for the scale parameter of a Weibull population with known shape parameter", *Technometrics* **7**, 405–422.

———— (1966), "Local maximum likelihood estimation of the parameters of three-parameter lognormal populations from complete and censored samples", *J. Amer. Statist. Assoc.* **61**, 842–851.

———— (1967), "Asymptotic variances and covariances of maximum likelihood estimators from censored samples, of parameters of Weibull and gamma distributions", *Ann. Math. Statist.* **38**, 557–563.

Hill, B. M. (1963), "The three parameter lognormal distribution and Bayesian analysis of a point-source epidemic", *J. Amer. Statist. Assoc.* **58**, 72–84.

Hogg, R. V. and Craig, A. T. (1978), *Introduction to Mathematical Statistics*, Macmillan, New York.

Jeffreys, H. (1961), *Theory of Probability*, Oxford University Press.

Johns, Jr., M. V. and Lieberman, G. J. (1966), "An exact asymptotically efficient bound for reliability in the case of the Weibull distribution", *Technometrics* **8**, 135–175.

Joshi, P. C. (1972), "Efficient estimation of the mean of an exponential distribution when an outlier is present", *Technometrics* **14**, 137–144.

Kale, B. K. (1962), "On the solution of the likelihood equations by iterative processes—multiparameter case", *Biometrika* **49**, 479–486.

Kale, B. K. and Sinha, S. K. (1971), "Estimation of expected life in the presence of an outlier observation", *Technometrics* **13**, 755–759.

Kao, J. H. K. (1959), "A graphical estimation of mixed Weibull parameters in life testing electron tubes", *Technometrics* **1**, 389–407.

Kendall, M. G. and Stuart, A. (1972), *Advanced Theory of Statistics—Vol. II*, Charles Griffin, London.

Kendall, P. J. (1963), "Estimation of the mean of the exponential distribution from grouped data when the sample is censored—with applications to life testing", Unpublished Ph.D. dissertation, North Carolina State University.

Kulldorff, G. (1962), "On the asymptotic optimum spacings for the estimation of the scale parameter of an exponential distribution based on sample quantiles", Mimeographed Report, Department of Statistics, University of Lund.

Larson, H. J. (1969), *Introduction to Probability Theory and Statistical Inference*, Wiley.

Laurent, A. G. (1962), "Conditional distribution of order statistics and distribution of the reduced *i*th order statistics of the exponential model", *Ann. Math. Statist.* **34**, 652–657.

Lawless, J. F.(1972), "On the estimation of safe life when the underlying life distribution is Weibull", *Technometrics* **15**, 857–864.

Lehmann, E. L. (1959), *Testing Statistical Hypotheses*, Wiley.

Lehmann, E. L. and Scheffé, Henry (1950, 1955), "Completeness, similar regions and unbiased estimation", *Sankhya* **10**, 305–340; **15**, 219–236.

Lentner, M. M. and Buehler, R. J. (1963), "Some inference about gamma parameters with applications to a reliability problem", *J. Amer. Statist. Assoc.* **58**, 670–677.

Lieberman, G. J. and Ross, S. M. (1971), "Confidence intervals for independent exponential series systems", *J. Amer. Statist. Assoc.* **66**, 837–840.

Lieblein, J. and Zelen, M. (1956), "Statistical investigation of the fatigue life of deep groove ball bearings", *J. of Res. Natl. Bur. Std.* **57**, 273–315.

Lindley, D. V. (1965), *Introduction to Probability and Statistics from a Bayesian Viewpoint*, Part 2, Inference, Cambridge University Press.

Littel, A. S. (1952), "Estimation of the T-year survival rate from follow-up studies over a limited period of time", *Human Biology* **24**, 87–116.

Maguire, B. A., Pearson, E. S. and Wynn, A. H. A. (1952), "The time intervals between industrial accidents", *Biometrika* **39**, 168–180.

Mann, N. (1968), "Results on statistical estimation and hypotheses testing with application to the Weibull and extreme value distribution", Aerospace Research Laboratories, Wright-Patterson Air Force Base, Ohio.

Mann, N., Schaffer, E. and Singpurwalla, N. (1974), *Methods for Statistical Analysis of Reliability and Life Data*, Wiley.

Mawaziny, E. and Buehler, R. J. (1967), "Confidence limits for the reliability of series systems", *J. Amer. Statist. Assoc.* **62**, 1452–1459.

Mendenhall, W. and Hader, R. J. (1958), "Estimation of parameters of mixed exponentially distributed failure time distributions from censored failure data", *Biometrika* **45**, 504–520.

Mendenhall, W. and Lehman, Jr., E. H. (1960), "An approximation to negative moments of positive binomial useful in life testing", *Technometrics* **2**, 227–242.

Mennon, M. V. (1963), "Estimation of the shape and scale parameters of the Weibull distribution", *Technometrics* **5**, 175–182.

Mood, A. M., Graybill, F. A. and Boes, D. C. (1974), *Introduction to the Theory of Statistics*, McGraw-Hill.

Moroney, M. J. (1951), *Facts from Figures*, Penguin, London.

Ogawa, J. (1960), "Determination of optimum spacings for the estimation of the scale parameter of an exponential distribution based on sample quantiles", *A. Inst. Statist. Math.* **2**, 135–141.

Pairman, E. (1919), *Tables of the Di-gamma and Tri-gamma Functions*, in Tracts for Computers No. 1, Karl Pearson (Ed.), Cambridge University Press.

Pearson, E. S. and Hartley, H. O. (1954), *Biometrika Tables for Statisticians, Vols. 1 and 2*, Cambridge University Press.

Pearson, Karl, *Tables of the Incomplete Beta Function*, Cambridge University Press (1968).
————— *Tables of the Incomplete Gamma Function*, Cambridge University Press (1946).

Pierce, D. A. (1973), "Fiducial, frequency and Bayesian inference on reliability for the two-parameter negative exponential distribution", *Technometrics* **15**, 249–253.

Plackett, R. L. (1959), "The analysis of life test data", *Technometrics* **1**, 9–19.

Pugh, E. L. (1963), "The best estimate of reliability with exponential case", *Op. Res.* **11**, 57–61.

Raiffa, H. and Schlaifer (1961), *Applied Statistical Decision Theory*, Harvard University Press.

Ranganathan, J. and Kale, B. K. (1978), "Tests of Hypotheses for Reliability function in Two Parameter Exponential Model", *Technical Report No. 85*, Department of Statistics, University of Manitoba, Winnipeg, Canada.

Resinikoff, G. J. and Lieberman, G. J. (1957), *Tables of Non-central t Distribution*, Stanford University Press.

Rider, P. R. (1961), "The method of moments applied to a mixture of two exponential distributions", *Ann. Math. Statist.* **32**, 143–147.

Rohatgi, V. K. (1976), *An Introduction to Probability Theory and Mathematical Statistics*, Wiley.

Rutemiller, H. C. (1966), "Point estimation of reliability of a system comprised of k elements from the same exponential distribution", *J. Amer. Statist. Assoc.* **61**, 1029–1032.

Saleh, A. K. M. E. (1964), "On the estimation of parameters of the exponential distribution based on optimum order statistics in censored samples", Ph.D. dissertation, University of Western Ontario.

Saleh, A. K. M. E. and Ali, M. M. (1966), "Asymptotic optimum quantiles for the estimation of the parameters of the negative exponential distribution", *Ann. Math. Statist.* **37**, 143–151.

Samanta, M. and Sinha, S. K. (1979), "Efficient estimation of reliability of two-component systems".

Sarndal, C. (1962), *Information from Censored Samples*, Almqvist and Wiksell, Stockholm.

Sarhan, A. and Greenberg, B. C. (1962), *Contributions to Order Statistics*, Wiley.

Sarkar, T. K. (1971), "An exact lower confidence bound for the reliability of a series system where each component has an exponential time to failure distribution", *Technometrics* **13**, 335–346.

Savage, L. J. (1962), *The Foundations of Statistical Inference: a Discussion*, Methuen, London.

Siddiqui, M. M. (1963), "Optimum estimators of the parameters of negative exponential distribution from one or two order statistics", *Ann. Math. Statist.* **34**, 117–121.

Sinha, S. K. (1972), "Reliability estimation in life testing in the presence of an outlier observation", *Op. Res.* **20**, 888–894.

———— (1979), "Robustness of posteriors under improper priors in some life testing models", To appear in the Journal of the Indian Statistical Association.

Sinha, S. K. and Guttman, I. (1976), "Bayesian inference about the reliability function for exponential distributions", *Commun. Statist.—Theor. Meth.* A5(5), 471–479.

Sinha, S. K. and Lee, H. K. (1979), "Estimation of mixed Weibull parameters with complete and censored data", *Indian Association for Productivity, Quality and Reliability Transactions* **3** (1).

Smith, W. L. (1957), "A note on truncation and sufficient statistics", *Ann. Math. Statist.* **30**, 341–366.

Swamy, P. S. and Doss, A. D. C. (1961), "On a problem of Bartholomew in life testing", *Sankhya* A, **23** (3), 225–230.

Tate, R. F. (1959), "Unbiased estimation: Functions of location and scale parameters", *Ann. Math. Statist.* **30**, 341–366.

Tate, R. F. and Klett, G. W. (1959), "Optimum confidence interval for the variance of a normal distribution", *J. Amer. Statist. Assoc.* **54**, 674–682.

Teichrow, D. (1962), *Contributions to Order Statistics*, Wiley 190–205.

Thoman, D. R., Bain, L. J. and Antle, C. E. (1969), "Inferences on the parameters of the Weibull distribution", *Technometrics* **11**, 445–460.

Tiku, M. L. (1968), "Estimating the parameters of lognormal distribution from censored samples", *J. Amer. Statist. Assoc.* **63**, 134–140.

Tukey, J. W. (1949), "Sufficiency, truncation and selections", *Ann. Math. Statist.* **20**, 309–311.

Wani, J. K. and Kabe, D. G. (1971), "Point estimation of reliability of a system comprised of k elements from the same gamma distribution", *Technometrics* **13**, 859–864.

Watson, G. N. (1952), *Treatise on the theory of Bessel Functions* (2nd ed.), Cambridge University Press.

Weibull, W. (1939), "A statistical theory of the strength of materials", *Ing. Vetenskaps Akad. Handl.* **151**, 1–45.

———— (1951), "A statistical distribution function of wide applicability", *J. Appl. Mech.* **18**, 293–297.

White, J. S. (1969), "The moments of Log-Weibull order statistics", *Technometrics* **11**, 373–386.

Wilks, S. S. (1969), *Mathematical Statistics*, Wiley.

Wilk, M. B., Gnanadesikan, R. and Huyette, M. J. (1962), "Estimation of parameters of the gamma distribution using order statistics", *Biometrika* **49**, 525–545.

Zacks, S. and Even, M. (1966), "The efficiencies in small samples of the maximum likelihood and best unbiased estimators of reliability functions", *J. Amer. Statist. Assoc.* **61**, 1033–1050.

Zacks, S. (1971), *The Theory of Statistical Inference*, Wiley.

Zehna, P. W. (1966), "Invariance of maximum likelihood estimators", *Ann. Math. Statist.* **66**, 891.

Index